対訳

ISO/IEC 17025:2017

(JIS Q 17025:2018)

ポケット版

試験所及び校正機関の能力 に関する一般要求事項

日本規格協会 編

＊著作権について

本書は，ISO 中央事務局と当会との翻訳出版契約に基づいて刊行したものです．

本書に収録した ISO 及び JIS は，著作権により保護されています．本書の一部又は全部について，当会及び ISO の許可なく複写・複製することを禁じます．ISO の著作権は，下に示すとおりです．

本書の著作権に関するお問合せは，当会販売サービスチーム（Tel. 03-4231-8550）にて承ります．

©ISO/IEC 2017, Published in Switzerland

All rights reserved. Unless otherwise specified, no part of this publication may be reproduced or utilized otherwise in any form or by any means, electronic or mechanical, including photocopying, or posting on the internet or an intranet, without prior written permission. Permission can be requested from either ISO at the address below or ISO's member body in the country of the requester.

ISO copyright office
Ch. de Blandonnet 8 · CP 401
CH-1214 Vernier, Geneva, Switzerland
Tel. + 41 22 749 01 11
Fax + 41 22 749 09 47
copyright@iso.org
www.iso.org

本書について

本書は，国際標準化機構（ISO）が2017年11月に第3版として発行した国際規格 ISO/IEC 17025: 2017（General requirements for the competence of testing and calibration laboratories），及びその翻訳規格として，日本工業標準調査会（JISC）の審議を経て2018年7月20日に経済産業大臣が改正した日本工業規格 JIS Q 17025:2018（試験所及び校正機関の能力に関する一般要求事項）を，英和対訳で収録したものです．

収録に際して JIS の解説は省略しています．JIS の解説を参照したい場合は，JIS 規格票（JIS Q 17025: 2018）をご利用ください．

また，これらの規格をより深く理解したい方には，書籍『ISO/IEC 17025:2017（JIS Q 17025:2018）要求事項の解説（仮称）』（当会より2018年11月末発行予定）を併読されることをお勧めします．

2018年10月

日本規格協会

JIS Q 17025 原案作成委員会

(委員長)	藤　間　一　郎	国立研究開発法人産業技術総合研究所
(委員)	浅　野　浩　太	一般財団法人日本品質保証機構
	井　口　新　一	一般社団法人 RMA
	植　松　慶　生	公益財団法人日本適合性認定協会
	榎　本　克　哉	公益社団法人産業安全技術協会
	大　高　広　明	独立行政法人製品評価技術基盤機構
	岸　本　勇　夫	国立研究開発法人産業技術総合研究所
	木　下　定　之	一般社団法人日本電機工業会
	菅　原　　　昇	一般社団法人日本環境測定分析協会
	塚　田　年　行	一般財団法人カケンテストセンター
(関係者)	人　村　朋　之	総務省
	中　井　清　人	厚生労働省
	高　橋　悠　一	厚生労働省
	阪　本　和　広	農林水産省
	峯戸松　勝　秀	農林水産省
	関　野　武　志	経済産業省
	荒　井　　　淳	経済産業省
	松　下　一　徳	経済産業省
	四ツ釜　直　哉	経済産業省
(事務局)	中　川　　　梓	一般財団法人日本規格協会
	千　葉　祐　介	一般財団法人日本規格協会

Contents

ISO/IEC 17025:2017
General requirements for the competence of testing and calibration laboratories

Foreword ·· 12

Introduction ·· 18

1 **Scope** ·· 22

2 **Normative references** ························· 24

3 **Terms and definitions** ······················ 24

4 **General requirements** ······················· 36

4.1 Impartiality ·· 36

4.2 Confidentiality ·· 40

5 **Structural requirements** ··················· 42

6 **Resource requirements** ···················· 48

6.1 General ·· 48

6.2 Personnel ·· 50

6.3 Facilities and environmental conditions
·· 52

6.4 Equipment ·· 56

6.5 Metrological traceability ························· 66

6.6 Externally provided products and services
·· 70

目　次

JIS Q 17025:2018
試験所及び校正機関の能力に関する
一般要求事項

まえがき ………………………………………………	13
序文 ……………………………………………………	19
1　適用範囲 ………………………………………	23
2　引用規格 ………………………………………	25
3　用語及び定義 …………………………………	25
4　一般要求事項 …………………………………	37
4.1　公平性 …………………………………………	37
4.2　機密保持 ………………………………………	41
5　組織構成に関する要求事項 …………………	43
6　資源に関する要求事項 ………………………	49
6.1　一般 ……………………………………………	49
6.2　要員 ……………………………………………	51
6.3　施設及び環境条件 ……………………………	53
6.4　設備 ……………………………………………	57
6.5　計量トレーサビリティ ………………………	67
6.6　外部から提供される製品及びサービス ……	71

7　Process requirements ··············· 76

7.1　Review of requests, tenders and contracts

··· 76

7.2　Selection, verification and validation

　　of methods ······························· 82

7.3　Sampling ································· 94

7.4　Handling of test or calibration items

··· 98

7.5　Technical records ···················· 100

7.6　Evaluation of measurement uncertainty

··· 102

7.7　Ensuring the validity of results ·········· 106

7.8　Reporting of results ···················· 110

7.9　Complaints ····························· 132

7.10　Nonconforming work ················· 136

7.11　Control of data and information

　　management ··························· 138

8　Management system requirements ····· 144

8.1　Options ································· 144

8.2　Management system documentation

　　(Option A) ····························· 146

8.3　Control of management system

　　documents (Option A) ················· 148

8.4　Control of records (Option A) ··········· 152

7 プロセスに関する要求事項 ·········· 77

7.1 依頼，見積仕様書及び契約のレビュー ······ 77

7.2 方法の選定，検証及び妥当性確認 ········· 83

7.3 サンプリング ································· 95

7.4 試験・校正品目の取扱い ···················· 99

7.5 技術的記録 ································· 101

7.6 測定不確かさの評価 ······················· 103

7.7 結果の妥当性の確保 ······················· 107

7.8 結果の報告 ································· 111

7.9 苦情 ····································· 133

7.10 不適合業務 ······························· 137

7.11 データの管理及び情報マネジメント ······ 139

8 マネジメントシステムに関する要求事項 ······ 145

8.1 選択肢 ··································· 145

8.2 マネジメントシステムの文書化
（選択肢 A） ······························· 147

8.3 マネジメントシステム文書の管理
（選択肢 A） ······························· 110

8.4 記録の管理（選択肢 A） ···················· 153

10

8.5 Actions to address risks and
opportunities (Option A) ·····················152

8.6 Improvement (Option A) ·····················156

8.7 Corrective actions (Option A) ···············158

8.8 Internal audits (Option A) ····················162

8.9 Management reviews (Option A) ···········164

Annex A (informative) **Metrological traceability**
···170

Annex B (informative) **Management system options** ···180

Bibliography ···186

8.5 リスク及び機会への取組み（選択肢 A）… 153

8.6 改善（選択肢 A）……………………… 157

8.7 是正処置（選択肢 A）………………… 159

8.8 内部監査（選択肢 A）………………… 163

8.9 マネジメントレビュー（選択肢 A）……… 165

附属書 A（参考）　計量トレーサビリティ……… 171

附属書 B（参考）　マネジメントシステムに関する選択肢……………………………… 181

参考文献 ……………………………………… 187

Foreword

ISO (the International Organization for Standardization) is a worldwide federation of national standards bodies (ISO member bodies). The work of preparing International Standards is normally carried out through ISO technical committees. Each member body interested in a subject for which a technical committee has been established has the right to be represented on that committee. International organizations, governmental and non-governmental, in liaison with ISO, also take part in the work. In the field of conformity assessment, ISO and the International Electrotechnical Commission (IEC) develop joint ISO/IEC documents under the management of the ISO Committee on Conformity assessment (ISO/CASCO).

The procedures used to develop this document and those intended for its further maintenance are described in the ISO/IEC Directives, Part 1. In particular the different approval criteria needed

まえがき

(ISO の Foreword と JIS のまえがきは，それぞれの原文において内容が異なっているため，対訳となっていないことにご注意ください．)

この規格は，工業標準化法に基づき，日本工業標準調査会の審議を経て，経済産業大臣が改正した日本工業規格である．

これによって，**JIS Q 17025**:2005 は改正され，この規格に置き換えられた．

この規格は，著作権法で保護対象となっている著作物である．

この規格の一部が，特許権，出願公開後の特許出願又は実用新案権に抵触する可能性があることに注意を喚起する．経済産業大臣及び日本工業標準調査会は，このような特許権，出願公開後の特許出願及び実用新案権に関わる確認について，責任はもたない．

14 ISO/IEC 17025

for the different types of ISO documents should be noted. This document was drafted in accordance with the editorial rules of the ISO/IEC Directives, Part 2 (see **www.iso.org/directives**).

Attention is drawn to the possibility that some of the elements of this document may be the subject of patent rights. ISO shall not be held responsible for identifying any or all such patent rights. Details of any patent rights identified during the development of the document will be in the Introduction and/or on the ISO list of patent declarations received (see **www.iso.org/patents**).

Any trade name used in this document is information given for the convenience of users and does not constitute an endorsement.

For an explanation on the voluntary nature of standards, the meaning of ISO specific terms and expressions related to conformity assessment, as well as information about ISO's adherence to the World Trade Organization (WTO) principles in the Technical Barriers to Trade (TBT) see the follow-

15

16 ISO/IEC 17025

ing URL: **www.iso.org/iso/foreword.html**.

This document was prepared by the ISO Committee on Conformity Assessment (CASCO) and circulated for voting to the national bodies of both ISO and IEC, and was approved by both organizations.

This third edition cancels and replaces the second edition (ISO/IEC 17025:2005), which has been technically revised.

The main changes compared to the previous edition are as follows:
— the risk-based thinking applied in this edition has enabled some reduction in prescriptive requirements and their replacement by performance-based requirements;
— there is greater flexibility than in the previous edition in the requirements for processes, procedures, documented information and organizational responsibilities;
— a definition of "laboratory" has been added (see **3.6**).

17

Introduction

This document has been developed with the objective of promoting confidence in the operation of laboratories. This document contains requirements for laboratories to enable them to demonstrate they operate competently, and are able to generate valid results. Laboratories that conform to this document will also operate generally in accordance with the principles of ISO 9001.

This document requires the laboratory to plan and implement actions to address risks and opportunities. Addressing both risks and opportunities establishes a basis for increasing the effectiveness of the management system, achieving improved results and preventing negative effects. The laboratory is responsible for deciding which risks and opportunities need to be addressed.

序文

(ISO の Introduction と JIS の序文は，それぞれの原文におい
て内容が異なっているため，対訳となっていないことにご注意
ください．)

この規格は，2017 年に第 3 版として発行された
ISO/IEC 17025 を基に，技術的内容及び構成を変
更することなく作成した日本工業規格である．

この規格は，ラボラトリの運営の信頼性を高める
という目的をもって作成された．

この規格は，ラボラトリが適格な運営を行い，か
つ，妥当な結果を出す能力があることを実証できる
ようにするための要求事項を含んでいる．

この規格に適合するラボラトリは，一般に **JIS
Q 9001** の原則にも従った運営をすることになる．

この規格は，リスク及び機会に取り組むための処
置を計画し，実施することをラボラトリに要求して
いる．リスク及び機会の双方に取り組むことによっ
て，マネジメントシステムの有効性の向上，改善さ
れた結果の達成及び好ましくない影響の防止のため
の基礎が確立される．ラボラトリは，どのリスク及
び機会に取り組む必要があるかを決定する責任をも
つ．

The use of this document will facilitate cooperation between laboratories and other bodies, and assist in the exchange of information and experience, and in the harmonization of standards and procedures. The acceptance of results between countries is facilitated if laboratories conform to this document.

In this document, the following verbal forms are used:

— "shall" indicates a requirement;

— "should" indicates a recommendation;

— "may" indicates a permission;

— "can" indicates a possibility or a capability.

Further details can be found in the ISO/IEC Directives, Part 2.

For the purposes of research, users are encouraged to share their views on this document and their priorities for changes to future editions. Click on the link below to take part in the online survey:

17025_ed3_usersurvey

この規格の使用は，ラボラトリとその他の機関との間の協力を容易にし，情報及び経験の交換並びに規格及び手順の整合化を支援するであろう．ラボラトリがこの規格に適合している場合には，国家間での結果の受入れが容易になる．

1 Scope

This document specifies the general requirements for the competence, impartiality and consistent operation of laboratories.

This document is applicable to all organizations performing laboratory activities, regardless of the number of personnel.

Laboratory customers, regulatory authorities, organizations and schemes using peer-assessment, accreditation bodies, and others use this document in confirming or recognizing the competence of laboratories.

1 適用範囲

この規格は，ラボラトリの能力，公平性及び一貫した運営に関する一般要求事項を規定する．

この規格は，要員の数に関係なく，ラボラトリ活動を行う全ての組織に適用できる．

ラボラトリの顧客，規制当局，相互評価を使用する組織及びスキーム並びに認定機関及びその他の組織が，ラボラトリの能力を確認又は承認するに当たってこの規格を使用する．

注記　この規格の対応国際規格及びその対応の程度を表す記号を，次に示す．

ISO/IEC 17025:2017, General requirements for the competence of testing and calibration laboratories (IDT)

なお，対応の程度を表す記号"IDT"は，ISO/IEC Guide 21-1 に基づき，"一致している"ことを示す．

2 Normative references

The following documents are referred to in the text in such a way that some or all of their content constitutes requirements of this document. For dated references, only the edition cited applies. For undated references, the latest edition of the referenced document (including any amendments) applies.

ISO/IEC Guide 99, *International vocabulary of metrology — Basic and general concepts and associated terms (VIM)*[1]

ISO/IEC 17000, *Conformity assessment — Vocabulary and general principles*

3 Terms and definitions

For the purposes of this document, the terms and definitions given in ISO/IEC Guide 99 and ISO/IEC 17000 and the following apply.

---

1) Also known as JCGM 200.

2 引用規格

次に掲げる規格は，この規格に引用されることによって，この規格の規定の一部を構成する．これらの引用規格は，その最新版（追補を含む．）を適用する．

JIS Q 17000　適合性評価－用語及び一般原則
　　注記　対応国際規格：ISO/IEC 17000, Conformity assessment — Vocabulary and general principles (IDT)
　　　　　ISO/IEC Guide 99, International vocabulary of metrology — Basic and general concepts and associated terms (VIM)[1]
　　注[1]　JCGM 200 としても知られている．

3 用語及び定義

この規格で用いる主な用語及び定義は，ISO/IEC Guide 99 及び JIS Q 17000 によるほか，次による．

26 ISO/IEC 17025

ISO and IEC maintain terminological databases for use in standardization at the following addresses:

— ISO Online browsing platform: available at https://www.iso.org/obp

— IEC Electropedia: available at http://www.electropedia.org/

3.1

impartiality

presence of objectivity

Note 1 to entry: Objectivity means that conflicts of interest do not exist, or are resolved so as not to adversely inf luence subsequent activities of the *laboratory* (**3.6**).

Note 2 to entry: Other terms that are useful in conveying the element of impartiality include "freedom from conflict of interests", "freedom from bias", "lack of prejudice", "neutrality", "fairness", "open-mindedness", "even-handedness", "detachment", "balance".

[SOURCE: ISO/IEC 17021-1:2015, 3.2, modified

3　用語及び定義

3.1
公平性（impartiality）

客観性があること．

注記 1　客観性とは，利害抵触がないか，又は
ラボラトリの事後の活動に悪影響を及
ぼすことがないよう，利害抵触が解決
されていることを意味する．

注記 2　公平性の要素を伝えるのに有用なその
他の用語には，利害抵触がないこと，
偏見がないこと，先入観がないこと，
中立，公正，心が広いこと，公明正
大，利害との分離，及び均衡が含まれ
る．

（**出典：JIS Q 17021-1:2015，3.2** の**注記 1** にあ

— The words "the certification body" have been re-placed by "the laboratory" in Note 1 to entry, and the word "independence" has been deleted from the list in Note 2 to entry.]

3.2
complaint

expression of dissatisfaction by any person or organization to a *laboratory* (**3.6**), relating to the activities or results of that laboratory, where a response is expected

[SOURCE: ISO/IEC 17000:2004, 6.5, modified — The words "other than appeal" have been deleted, and the words "a conformity assessment body or accreditation body, relating to the activities of that body" have been replaced by "a laboratory, relating to the activities or results of that laboratory".]

3.3
interlaboratory comparison

organization, performance and evaluation of measurements or tests on the same or similar items by two or more laboratories in accordance with prede-

3.2

苦情（complaint）

ラボラトリの活動又は結果に関し，人又は組織が回答を期待して行う当該ラボラトリへの不満の表明．

（**出典：JIS Q 17000**:2005 の **6.5** を修正．"適合性評価機関又は認定機関" を "ラボラトリ" に置き換えた．また，"結果" を追加し，"異議申立て" を削除した．）

3.3

試験所間比較（interlaboratory comparison）

事前に定めた条件に従って，二つ以上のラボラトリが，同一品目又は類似品目で行う，測定又は試験の企画，実施及び評価．

30 ISO/IEC 17025

termined conditions

[SOURCE: ISO/IEC 17043:2010, 3.4]

3.4

intralaboratory comparison

organization, performance and evaluation of measurements or tests on the same or similar items within the same *laboratory* (**3.6**) in accordance with predetermined conditions

3.5

proficiency testing

evaluation of participant performance against preestablished criteria by means of *interlaboratory comparisons* (**3.3**)

[SOURCE: ISO/IEC 17043:2010, 3.7, modified — Notes to entry have been deleted.]

3.6

laboratory

body that performs one or more of the following activities:

(**出典：JIS Q 17043**:2011 の **3.4**)

3.4
試験所内比較（intralaboratory comparison）
　事前に定めた条件に従って，同一のラボラトリ内で，同一品目又は類似品目で行う，測定又は試験の企画，実施及び評価．

3.5
技能試験（proficiency testing）
　試験所間比較による，事前に決めた基準に照らしての参加者のパフォーマンスの評価．

（**出典：JIS Q 17043**:2011 の **3.7** を修正．**注記**を削除した．）

3.6
ラボラトリ（laboratory）
　次の　つ以上の活動を実行する機関．

32   ISO/IEC 17025

— testing;

— calibration;

— sampling, associated with subsequent testing or calibration

Note 1 to entry: In the context of this document, "laboratory activities" refer to the three above-mentioned activities.

3.7

decision rule

rule that describes how measurement uncertainty is accounted for when stating conformity with a specified requirement

3.8

verification

provision of objective evidence that a given item fulfils specified requirements

EXAMPLE 1   Confirmation that a given reference material as claimed is homogeneous for the quantity value and measurement procedure concerned, down to a measurement portion having a

— 試験

— 校正

— 後の試験又は校正に付随するサンプリング

> **注記** 現在の規格の枠組みにおいて，"ラボラ
> トリ活動"という用語は，上記三つの活
> 動のことをいう．

3.7
判定ルール（decision rule）

特定の要求事項への適合性を表明する際に，測定不確かさをどのように考慮するかを記述した取決め．

3.8
検証（verification）

与えられたアイテムが規定された要求事項を満たしているという客観的証拠の提示．

> **例1** 対象とする任意の標準物質が，当該の
> 量の値及び測定手順に対して，質量 10
> mg の測定試料まで均質であることの確
> 認．

34 ISO/IEC 17025

mass of 10 mg.

EXAMPLE 2 Confirmation that performance properties or legal requirements of a measuring system are achieved.

EXAMPLE 3 Confirmation that a target measurement uncertainty can be met.

Note 1 to entry: When applicable, measurement uncertainty should be taken into consideration.

Note 2 to entry: The item may be, for example, a process, measurement procedure, material, compound, or measuring system.

Note 3 to entry: The specified requirements may be, for example, that a manufacturer's specifications are met.

Note 4 to entry: Verification in legal metrology, as defined in VIML, and in conformity assessment in general, pertains to the examination and marking and/or issuing of a verification certificate for a measuring system.

Note 5 to entry: Verification should not be confused with calibration. Not every verification is a *validation* (**3.9**).

Note 6 to entry: In chemistry, verification of the

3 用語及び定義　　　35

例 2　測定システムが性能特性又は法的要求事項を満たしていることの確認

例 3　目標測定不確かさを満たすことができることの確認.

注記 1　適用可能な場合は，測定不確かさを考慮することが望ましい.

注記 2　アイテムとは，例えば，プロセス，測定手順，材料，化合物又は測定システムのいずれであってもよい.

注記 3　規定された要求事項とは，例えば，製造業者の仕様を満たしていることである.

注記 4　VIML 及び一般に適合性評価で定義しているように，法定計量でいう検証は，評価及び表示，及び／又は測定システムに対する検定証明書の発行を含む.

注記 5　検証と校正とを混同しないようにすることが望ましい. 全ての検証が妥当性確認であるとは限らない.

注記 6　化学の分野では，関連する実在物又は

36 ISO/IEC 17025

identity of the entity involved, or of activity, requires a description of the structure or properties of that entity or activity.

[SOURCE: ISO/IEC Guide 99:2007, 2.44]

3.9
validation
verification (**3.8**), where the specified requirements are adequate for an intended use

EXAMPLE   A measurement procedure, ordinarily used for the measurement of mass concentration of nitrogen in water, may be validated also for measurement of mass concentration of nitrogen in human serum.

[SOURCE: ISO/IEC Guide 99:2007, 2.45]

4   General requirements
4.1   Impartiality
4.1.1   Laboratory activities shall be undertaken impartially and structured and managed so as to safeguard impartiality.

活性の同一性の検証には，その実在物若しくは活性の構造又は性質の記述が必要となる.

（出典：ISO/IEC Guide 99 の 2.44）

3.9

妥当性確認（validation）
規定された要求事項が意図した用途に十分であることの検証.

> **例** 水中の窒素の質量濃度の測定に通常用いる測定手順は，ヒトの血清中の窒素の質量濃度の測定に対しても妥当である場合がある.

（出典：ISO/IEC Guide 99 の 2.45）

4 一般要求事項

4.1 公平性

4.1.1 ラボラトリ活動は，公平に実行され，公平性を確保するように編成及び運営されなければならない.

4.1.2 The laboratory management shall be committed to impartiality.

4.1.3 The laboratory shall be responsible for the impartiality of its laboratory activities and shall not allow commercial, financial or other pressures to compromise impartiality.

4.1.4 The laboratory shall identify risks to its impartiality on an on-going basis. This shall include those risks that arise from its activities, or from its relationships, or from the relationships of its personnel. However, such relationships do not necessarily present a laboratory with a risk to impartiality.

NOTE   A relationship that threatens the impartiality of the laboratory can be based on ownership, governance, management, personnel, shared resources, finances, contracts, marketing (including branding), and payment of a sales commission or other inducement for the referral of new customers, etc.

4.1.2 ラボラトリマネジメントは，公平性を確約しなければならない．

4.1.3 ラボラトリは，ラボラトリ活動の公平性に対して責任をもたなければならず，公平性を損なう商業的，財務的，又はその他の圧力を容認してはならない．

4.1.4 ラボラトリは，公平性に対するリスクを継続的に特定しなければならない．ラボラトリの活動若しくは他との関係，又はその要員の他との関係をもつことから生じるリスクもこれに含めなければならない．ただし，そのような関係が，ラボラトリにとって必ずしも公平性に対するリスクになるとは限らない．

注記 ラボラトリの公平性に対する脅威となる関係としては，所有，統治，マネジメント，要員，共有資源，財務，契約，マーケティング（ブランド設定を含む．），及び新規顧客の紹介に関わる売上手数料の支払い又はその他の誘引条件に基づくものが挙げられる．

4.1.5 If a risk to impartiality is identified, the laboratory shall be able to demonstrate how it eliminates or minimizes such risk.

4.2 Confidentiality

4.2.1 The laboratory shall be responsible, through legally enforceable commitments, for the management of all information obtained or created during the performance of laboratory activities. The laboratory shall inform the customer in advance, of the information it intends to place in the public domain. Except for information that the customer makes publicly available, or when agreed between the laboratory and the customer (e.g. for the purpose of responding to complaints), all other information is considered proprietary information and shall be regarded as confidential.

4.2.2 When the laboratory is required by law or authorized by contractual arrangements to release confidential information, the customer or individual concerned shall, unless prohibited by law, be notified of the information provided.

4.1.5 公平性に対するリスクが特定された場合，ラボラトリは，そのリスクをどのように排除又は最小化するかを実証できなければならない．

4.2 機密保持

4.2.1 ラボラトリは，法的に強制力のあるコミットメントによって，ラボラトリ活動を実行する過程で得られた又は作成された全ての情報の管理について責任をもたなければならない．ラボラトリは，公開対象にしようとしている情報を，事前に顧客に通知しなければならない．顧客が公開している情報，又はラボラトリと顧客とが合意している場合（例えば，苦情への対応の目的のため）を除き，その他全ての情報は占有情報とみなし，機密としなければならない．

4.2.2 ラボラトリが機密情報を公開することを，法律で要求されるか又は契約上の取決めで認められた場合，顧客又は関係する個人は，法律によって禁止されない限り，当該情報の提供について知らされなければならない．

4.2.3 Information about the customer obtained from sources other than the customer (e.g. complainant, regulators) shall be confidential between the customer and the laboratory. The provider (source) of this information shall be confidential to the laboratory and shall not be shared with the customer, unless agreed by the source.

4.2.4 Personnel, including any committee members, contractors, personnel of external bodies, or individuals acting on the laboratory's behalf, shall keep confidential all information obtained or created during the performance of laboratory activities, except as required by law.

5   Structural requirements

5.1   The laboratory shall be a legal entity, or a defined part of a legal entity, that is legally responsible for its laboratory activities.

NOTE   For the purposes of this document, a governmental laboratory is deemed to be a legal entity on the basis of its governmental status.

4.2.3 当該顧客以外の情報源（例えば，苦情申立者，規制当局）から得られた顧客に関する情報は，顧客とラボラトリとの間で機密としなければならない．この情報の提供者（情報源）は，ラボラトリの機密とし，情報源が同意した場合を除き，顧客と共有してはならない．

4.2.4 委員会のメンバー，契約人，外部機関の要員又はラボラトリの代理人として活動する個人は，法律で要求される場合を除き，ラボラトリ活動を遂行する間に得られた，又は生じた全ての情報について機密保持しなければならない．

5　組織構成に関する要求事項

5.1　ラボラトリは，そのラボラトリ活動に法的責任をもつ法人であるか，又は法人の一部として明確に位置付けられていなければならない．

注記　この規格の目的において，政府のラボラトリは，政府機関としての地位に基づき法人とみなす．

5.2 The laboratory shall identify management that has overall responsibility for the laboratory.

5.3 The laboratory shall define and document the range of laboratory activities for which it conforms with this document. The laboratory shall only claim conformity with this document for this range of laboratory activities, which excludes externally provided laboratory activities on an ongoing basis.

5.4 Laboratory activities shall be carried out in such a way as to meet the requirements of this document, the laboratory's customers, regulatory authorities and organizations providing recognition. This shall include laboratory activities performed in all its permanent facilities, at sites away from its permanent facilities, in associated temporary or mobile facilities or at a customer's facility.

5.5 The laboratory shall:

5.2 ラボラトリは，そのラボラトリについて総合的な責任をもつラボラトリマネジメントを特定しなければならない．

5.3 ラボラトリは，この規格に適合するラボラトリ活動の範囲を明確化し，文書化しなければならない．ラボラトリは，継続的に外部から提供されるラボラトリ活動を除いた当該ラボラトリ活動の範囲に関してだけ，この規格への適合を主張しなければならない．

5.4 ラボラトリ活動は，この規格，ラボラトリの顧客，規制当局及び認可を与える機関の要求事項を満足するように実施されなければならない．このことには，その全ての恒久的施設で実施されるラボラトリ活動，その恒久的施設から離れた場所で実施されるラボラトリ活動，関連する一時施設若しくは移動施設で実施されるラボラトリ活動，又は顧客の施設で実施されるラボラトリ活動が含まれなければならない．

5.5 ラボラトリは，次の事項を行わなければならない．

46 ISO/IEC 17025

a) define the organization and management structure of the laboratory, its place in any parent organization, and the relationships between management, technical operations and support services;

b) specify the responsibility, authority and interrelationship of all personnel who manage, perform or verify work affecting the results of laboratory activities;

c) document its procedures to the extent necessary to ensure the consistent application of its laboratory activities and the validity of the results.

5.6 The laboratory shall have personnel who, irrespective of other responsibilities, have the authority and resources needed to carry out their duties, including:

a) implementation, maintenance and improvement of the management system;

b) identification of deviations from the management system or from the procedures for performing laboratory activities;

c) initiation of actions to prevent or minimize

a) ラボラトリの組織及び管理構造，親組織における位置付け，並びに管理，技術的職務及び支援サービスの間の関係を明確にする．

b) ラボラトリ活動の結果に影響する業務を管理，実施又は検証する全ての要員の責任，権限及び相互関係を規定する．

c) ラボラトリ活動の一貫した適用及び結果の妥当性を確実にするために必要な程度まで手順を文書化する．

5.6 ラボラトリは，他の責任のいかんにかかわらず，次の事項を含む責務を果たすために必要な権限及び資源をもつ要員をもたなければならない．

a) マネジメントシステムの実施，維持及び改善

b) マネジメントシステムからの逸脱，又はラボラトリ活動の実施手順からの逸脱の特定

c) それらの逸脱を防止又は最小化する処置の開始

48   ISO/IEC 17025

such deviations;

d)   reporting to laboratory management on the performance of the management system and any need for improvement;

e)   ensuring the effectiveness of laboratory activities.

5.7   Laboratory management shall ensure that:

a)   communication takes place regarding the effectiveness of the management system and the importance of meeting customers' and other requirements;

b)   the integrity of the management system is maintained when changes to the management system are planned and implemented.

6   Resource requirements
6.1   General

The laboratory shall have available the personnel, facilities, equipment, systems and support services necessary to manage and perform its laboratory activities.

d) マネジメントシステムの実施状況及び改善の必要性に関するラボラトリマネジメントへの報告

e) ラボラトリ活動の有効性の確保

5.7 ラボラトリマネジメントは，次の事項を確実にしなければならない．

a) コミュニケーションが，マネジメントシステムの有効性，並びに顧客要求事項及びその他の要求事項を満たすことの重要性に関して行われている．

b) マネジメントシステムに対する変更が計画され実施された場合，マネジメントシステムの"全体として整っている状態"（integrity）が維持されている．

6 資源に関する要求事項

6.1 一般

ラボラトリは，ラボラトリ活動の管理及び実施に必要な要員，施設，設備，システム及び支援サービスを利用できるようにしなければならない．

6.2 Personnel

6.2.1 All personnel of the laboratory, either internal or external, that could influence the laboratory activities shall act impartially, be competent and work in accordance with the laboratory's management system.

6.2.2 The laboratory shall document the competence requirements for each function influencing the results of laboratory activities, including requirements for education, qualification, training, technical knowledge, skills and experience.

6.2.3 The laboratory shall ensure that the personnel have the competence to perform laboratory activities for which they are responsible and to evaluate the significance of deviations.

6.2.4 The management of the laboratory shall communicate to personnel their duties, responsibilities and authorities.

6.2.5 The laboratory shall have procedure(s) and retain records for:

6.2 要員

6.2.1 ラボラトリ活動に影響を与え得る，ラボラトリの内部又は外部の全ての要員は，公平に行動し，力量をもち，ラボラトリのマネジメントシステムに従って業務を行わなければならない．

6.2.2 ラボラトリは，学歴，資格，教育・訓練，技術的知識，技能及び経験に関する要求事項を含め，ラボラトリ活動の結果に影響を与える各職務に関する力量要求事項を文書化しなければならない．

6.2.3 ラボラトリは，その要員が，責任をもつラボラトリ活動を実施し，かつ，逸脱の重大性を評価する力量をもつことを確実にしなければならない．

6.2.4 ラボラトリの管理要員は，要員に責務，責任及び権限を伝達しなければならない．

6.2.5 ラボラトリは，次の事項に関する手順をもち，記録を保持しなければならない．

a) determining the competence requirements;

b) selection of personnel;

c) training of personnel;

d) supervision of personnel;

e) authorization of personnel;

f) monitoring competence of personnel.

6.2.6 The laboratory shall authorize personnel to perform specific laboratory activities, including but not limited to, the following:

a) development, modification, verification and validation of methods;

b) analysis of results, including statements of conformity or opinions and interpretations;

c) report, review and authorization of results.

6.3 Facilities and environmental conditions

6.3.1 The facilities and environmental conditions shall be suitable for the laboratory activities and shall not adversely affect the validity of results.

NOTE Influences that can adversely affect the validity of results can include, but are not limited to, microbial contamination, dust, electromagnetic

6 資源に関する要求事項　　53

a) 力量要求事項の決定

b) 要員の選定

c) 要員の教育・訓練

d) 要員の監督

e) 要員への権限付与

f) 要員の力量の監視

6.2.6 ラボラトリは，特定のラボラトリ活動（次を含むが，これらに限定されない）を実施する権限を，要員に与えなければならない．

a) 方法の開発，変更，検証及び妥当性確認

b) 適合性の表明又は意見及び解釈を含めた，結果の分析

c) 結果の報告，レビュー及び承認

6.3　施設及び環境条件

6.3.1 施設及び環境条件は，ラボラトリ活動に適するものでなければならない．また，結果の妥当性に悪影響を及ぼしてはならない．

> **注記**　結果の妥当性に悪影響を及ぼし得る影響には，微生物学的汚染，ほこり，電磁障害，放射線，湿度，電力供給，温度，騒

disturbances, radiation, humidity, electrical supply, temperature, sound and vibration.

6.3.2   The requirements for facilities and environmental conditions necessary for the performance of the laboratory activities shall be documented.

6.3.3   The laboratory shall monitor, control and record environmental conditions in accordance with relevant specifications, methods or procedures or where they influence the validity of the results.

6.3.4   Measures to control facilities shall be implemented, monitored and periodically reviewed and shall include, but not be limited to:

a)   access to and use of areas affecting laboratory activities;

b)   prevention of contamination, interference or adverse inf luences on laboratory activities;

c)   effective separation between areas with incompatible laboratory activities.

音及び振動が含まれるが，これらに限定されない．

6.3.2 ラボラトリ活動の実施に必要な施設及び環境条件に関する要求事項を文書化しなければならない．

6.3.3 ラボラトリは，該当する仕様書，方法若しくは手順書に従い，又は環境条件が結果の妥当性に影響を及ぼす場合には，環境条件を監視し，制御し記録しなければならない．

6.3.4 施設を管理するための手段を実施し，監視し，定期的に見直さなければならない．これらの手段には，次の事項が含まれなければならないが，これらに限定されない．

a) ラボラトリ活動に影響を及ぼす区域への立入り及びこれらの区域の使用

b) 汚染，干渉又はラボラトリ活動への悪影響の防止

c) 両立不可能なラボラトリ活動が行われる区域間の効果的な分離

6.3.5 When the laboratory performs laboratory activities at sites or facilities outside its permanent control, it shall ensure that the requirements related to facilities and environmental conditions of this document are met.

6.4 Equipment

6.4.1 The laboratory shall have access to equipment (including, but not limited to, measuring instruments, software, measurement standards, reference materials, reference data, reagents, consumables or auxiliary apparatus) that is required for the correct performance of laboratory activities and that can influence the results.

NOTE 1 A multitude of names exist for reference materials and certified reference materials, including reference standards, calibration standards, standard reference materials and quality control materials. ISO 17034 contains additional information on reference material producers (RMPs). RMPs that meet the requirements of ISO 17034 are considered to be competent. Reference materials from RMPs meeting the requirements

6　資源に関する要求事項　　　57

6.3.5　ラボラトリが自身の恒久的な管理下にない場所又は施設でラボラトリ活動を実施する場合は，この規格の施設及び環境条件に関する要求事項が満たされることを確実にしなければならない．

6.4　設備

6.4.1　ラボラトリは，ラボラトリ活動の適正な実施に必要で，かつ，結果に影響を与え得る設備（これには測定装置，ソフトウェア，測定標準，標準物質，参照データ，試薬及び消耗品又は補助の器具を含むが，これらに限定されない）が利用可能でなければならない．

注記 1　標準物質及び認証標準物質（CRM）には，参照標準，校正用標準物質，参照標準物質（SRM），品質管理用物質を含め，多数の名称が存在する．**JIS Q 17034** は，標準物質生産者（RMP）に関する追加情報を含んでいる．**JIS Q 17034** の要求事項を満たす RMPは，能力があるとみなされる．**JIS Q 17034** の要求事項を満たす生産者か

of ISO 17034 are provided with a product information sheet/certificate that specifies, amongst other characteristics, homogeneity and stability for specified properties and, for certified reference materials, specified properties with certified values, their associated measurement uncertainty and metrological traceability.

NOTE 2 ISO Guide 33 provides guidance on the selection and use of reference materials. ISO Guide 80 provides guidance to produce in-house quality control materials.

6.4.2 When the laboratory uses equipment outside its permanent control, it shall ensure that the requirements for equipment of this document are met.

6.4.3 The laboratory shall have a procedure for handling, transport, storage, use and planned maintenance of equipment in order to ensure proper functioning and to prevent contamination or deterioration.

6 資源に関する要求事項　　　59

ら入手した標準物質には，製品情報シ
ート，認証書が添えられている．そこ
には，その他の特性とともに，規定特
性の均質性及び安定性が記載されてお
り，認証標準物質については，更に認
証値及び付随する測定不確かさ並びに
計量トレーサビリティをもつ規定特性
が記載されている．

注記 2　**JIS Q 0033** は，標準物質の選択及び
使用に関する手引を提供する．**ISO
Guide 80** は，内部で品質管理用物質
を生産するための手引を提供する．

6.4.2　ラボラトリが自身の恒久的な管理下にない
設備を使用する場合は，この規格の設備に関する要
求事項が満たされることを確実にしなければならな
い．

6.4.3　ラボラトリは，設備が適正に機能すること
を確実にするため及び汚染又は劣化を防止するため
に，設備の取扱い，輸送，保管，使用及び計画的保
守の手順をもたなければならない．

6.4.4 The laboratory shall verify that equipment conforms to specified requirements before being placed or returned into service.

6.4.5 The equipment used for measurement shall be capable of achieving the measurement accuracy and/or measurement uncertainty required to provide a valid result.

6.4.6 Measuring equipment shall be calibrated when:

— the measurement accuracy or measurement uncertainty affects the validity of the reported results, and/or

— calibration of the equipment is required to establish the metrological traceability of the reported results.

NOTE Types of equipment having an effect on the validity of the reported results can include:

— those used for the direct measurement of the measurand, e.g. use of a balance to perform a mass measurement;

— those used to make corrections to the mea-

6 資源に関する要求事項 61

6.4.4 ラボラトリは，設備を業務使用に導入する前又は業務使用に復帰させる前に，規定された要求事項への適合を検証しなければならない．

6.4.5 測定に使用される設備は，妥当な結果を得るために必要な測定の精確さ及び／又は測定不確かさを達成する能力をもたなければならない．

6.4.6 測定設備は，次の場合に校正されなければならない．

— 測定の精確さ又は測定不確かさが，報告された結果の妥当性に影響を与える．

— その設備の校正が，報告された結果の計量トレーサビリティを確立するために要求される．

注記　報告された結果の妥当性に影響を及ぼす設備には，次が含まれ得る．

— 測定対象量の直接測定に使用される設備．例えば，質量の測定を行うために，はかりを使用する場合．

— 測定値の補正に使用される設備．例

62   ISO/IEC 17025

sured value, e.g. temperature measurements;

— those used to obtain a measurement result calculated from multiple quantities.

6.4.7 The laboratory shall establish a calibration programme, which shall be reviewed and adjusted as necessary in order to maintain confidence in the status of calibration.

6.4.8 All equipment requiring calibration or which has a defined period of validity shall be labelled, coded or otherwise identified to allow the user of the equipment to readily identify the status of calibration or period of validity.

6.4.9 Equipment that has been subjected to overloading or mishandling, gives questionable results, or has been shown to be defective or outside specified requirements, shall be taken out of service. It shall be isolated to prevent its use or clearly labelled or marked as being out of service until it has been verified to perform correctly. The laboratory shall examine the effect of the defect or deviation from specified requirements and shall

えば，温度測定．

— 複数の量から計算された測定結果を
得るために使用される設備．

6.4.7 ラボラトリは，校正プログラムを確立しな
ければならない．その校正プログラムは，校正状態
についての信頼を維持するため，見直され，必要に
応じて調整されなければならない．

6.4.8 校正が必要な全ての設備又は有効期間が定
められた全ての設備は，設備の使用者が校正状態又
は有効期間を容易に識別できるように，ラベル付け
を行うか，コード化するか，又はその他の方法で識
別しなければならない．

6.4.9 過負荷又は誤った取扱いを受けた設備，疑
わしい結果を生じる設備，又は欠陥をもつ若しくは
規定の要求事項を満たさないことが認められた設備
は，業務使用を停止しなければならない．その設備
は，それが正常に機能することが検証されるまで，
使用を防止するため隔離するか，又は業務使用停止
中であることを示す明瞭なラベル付け若しくはマー
ク付けを行わなければならない．ラボラトリは，不
具合又は規定された要求事項からの逸脱の影響を調

64 ISO/IEC 17025

initiate the management of nonconforming work procedure (see **7.10**).

6.4.10 When intermediate checks are necessary to maintain confidence in the performance of the equipment, these checks shall be carried out according to a procedure.

6.4.11 When calibration and reference material data include reference values or correction factors, the laboratory shall ensure the reference values and correction factors are updated and implemented, as appropriate, to meet specified requirements.

6.4.12 The laboratory shall take practicable measures to prevent unintended adjustments of equipment from invalidating results.

6.4.13 Records shall be retained for equipment which can influence laboratory activities. The records shall include the following, where applicable:
a)   the identity of equipment, including software and firmware version;
b)   the manufacturer's name, type identification,

査し，不適合業務の管理の手順を開始しなければならない（7.10参照）．

6.4.10 設備の機能についての信頼を維持するために中間チェックが必要な場合には，これらのチェックは，手順に従って実施しなければならない．

6.4.11 校正及び標準物質データに参照値又は補正因子が含まれる場合，ラボラトリは，規定された要求事項を満たすために，必要に応じて，参照値及び補正因子が更新され，有効に使用されることを確実にしなければならない．

6.4.12 ラボラトリは，意図しない設備の調整によって結果が無効となることを防ぐために，実行可能な手段を講じなければならない．

6.4.13 ラボラトリ活動に影響を与え得る設備の記録を保持しなければならない．記録には，適用可能な場合，次の事項を含めなければならない．

a) ソフトウェア及びファームウェアのバージョンを含む，設備の識別．

b) 製造業者の名称，型式の識別及びシリアル番号

and serial number or other unique identification;

c) evidence of verification that equipment conforms with specified requirements;

d) the current location;

e) calibration dates, results of calibrations, adjustments, acceptance criteria, and the due
date of the next calibration or the calibration
interval;

f) documentation of reference materials, results,
acceptance criteria, relevant dates and the period of validity;

g) the maintenance plan and maintenance carried out to date, where relevant to the performance of the equipment;

h) details of any damage, malfunction, modification to, or repair of, the equipment.

6.5 Metrological traceability

6.5.1 The laboratory shall establish and maintain metrological traceability of its measurement
results by means of a documented unbroken chain
of calibrations, each contributing to the measurement uncertainty, linking them to an appropriate

6 資源に関する要求事項 67

又はその他の固有の識別.

c) 設備が規定された要求事項に適合していることの検証の証拠.

d) 現在の所在場所.

e) 校正の日付, 校正結果, 調整, 受入基準及び次回校正の期日又は校正周期.

f) 標準物質の文書, 結果, 受入基準, 関連する日付及び有効期間.

g) 設備の機能に関連する場合は, 保守計画及びこれまでに実施された保守.

h) 設備の損傷, 機能不良, 改造又は修理の詳細.

6.5 計量トレーサビリティ

6.5.1 ラボラトリは, 測定結果を適切な計量参照に結び付けるよう, それぞれの校正が測定不確かさに寄与している, 文書化された切れ目のない校正の連鎖によって, 測定結果の計量トレーサビリティを確立し, 維持しなければならない.

reference.

NOTE 1   In ISO/IEC Guide 99, metrological trace-ability is defined as the "property of a measure-ment result whereby the result can be related to a reference through a documented unbroken chain of calibrations, each contributing to the measure-ment uncertainty".

NOTE 2   See **Annex A** for additional information on metrological traceability.

6.5.2   The laboratory shall ensure that measure-ment results are traceable to the International System of Units (SI) through:

a)   calibration provided by a competent labora-tory; or

   NOTE 1   Laboratories fulfilling the require-ments of this document are considered to be competent.

b)   certified values of certified reference materi-als provided by a competent producer with stated metrological traceability to the SI; or

   NOTE 2   Reference material producers ful-

注記1 ISO/IEC Guide 99 には，計量トレーサビリティは，"それぞれが測定正確かさに寄与している，文書化された切れ目のない校正の連鎖によって計量参照に測定結果を関係付けることができるという測定結果の性質"として定義されている．

注記2 計量トレーサビリティに関する追加の情報については，**附属書A**を参照．

6.5.2 ラボラトリは，次のいずれかを通じて，測定結果が国際単位系（SI）にトレーサブルであることを確実にしなければならない．

a) 能力のあるラボラトリから提供される校正．

注記1 この規格の要求事項を満たすラボラトリは，能力があるとみなされる．

b) 能力のある生産者から提供された，表明されたSIへの計量トレーサビリティを伴った認証標準物質の認証値．

注記2 **JIS Q 17034** の要求事項を満たす

filling the requirements of ISO 17034 are considered to be competent.

c) direct realization of the SI units ensured by comparison, directly or indirectly, with national or international standards.

NOTE 3 Details of practical realization of the definitions of some important units are given in the SI brochure.

6.5.3 When metrological traceability to the SI units is not technically possible, the laboratory shall demonstrate metrological traceability to an appropriate reference, e.g.:

a) certified values of certified reference materials provided by a competent producer;

b) results of reference measurement procedures, specified methods or consensus standards that are clearly described and accepted as providing measurement results fit for their intended use and ensured by suitable comparison.

6.6 Externally provided products and services

標準物質生産者は，能力があるとみなされる．

c) 直接的に又は間接的に，国家標準又は国際標準との比較によって確認がなされた SI 単位の直接的実現．

> **注記 3** 幾つかの重要な単位の定義の現実的な実現方法の詳細は，SI 文書に記載されている．

6.5.3 SI 単位に対する計量トレーサビリティが技術的に不可能である場合，ラボラトリは，例えば，次のような適切な計量参照への計量トレーサビリティを実証しなければならない．

a) 能力のある生産者から提供された認証標準物質の認証値．

b) 明確に記述され，意図した用途に合致した測定結果を提供するものとして受け入れられており，適切な比較によって確認がなされた参照測定手順，規定された方法又は合意標準の結果．

6.6 外部から提供される製品及びサービス

6.6.1   The laboratory shall ensure that only suitable externally provided products and services that affect laboratory activities are used, when such products and services:

a)   are intended for incorporation into the laboratory's own activities;

b)   are provided, in part or in full, directly to the customer by the laboratory, as received from the external provider;

c)   are used to support the operation of the laboratory.

NOTE   Products can include, for example, measurement standards and equipment, auxiliary equipment, consumable materials and reference materials. Services can include, for example, calibration services, sampling services, testing services, facility and equipment maintenance services, proficiency testing services and assessment and auditing services.

6.6.2   The laboratory shall have a procedure and retain records for:

a)   defining, reviewing and approving the labo-

6　資源に関する要求事項　　73

6.6.1　ラボラトリは，ラボラトリ活動に影響を及ぼす，外部から提供される製品及びサービスが次の事項に該当する場合には，適切なものだけが使用されることを確実にしなければならない．

a)　製品及びサービスがラボラトリ自体の活動に組み込まれることを意図したものである場合．

b)　製品及びサービスの一部又は全てが，外部提供者から受領したままの状態でラボラトリから顧客に直接提供される場合．

c)　製品及びサービスが，ラボラトリの業務を支援するために使用される場合．

　　注記　製品には，例えば，測定標準並びに設備，補助設備，消耗品及び標準物質が含まれ得る．サービスには，例えば，校正サービス，サンプリングサービス，試験サービス，施設及び設備保守サービス，技能試験サービス並びに評価及び監査サービスが含まれ得る．

6.6.2　ラボラトリは，次の事項に関する手順をもち記録を保持しなければならない．

a)　外部から提供される製品及びサービスに関する

ratory's requirements for externally provided products and services;

b) defining the criteria for evaluation, selection, monitoring of performance and re-evaluation of the external providers;

c) ensuring that externally provided products and services conform to the laboratory's established requirements, or when applicable, to the relevant requirements of this document, before they are used or directly provided to the customer;

d) taking any actions arising from evaluations, monitoring of performance and re-evaluations of the external providers.

6.6.3 The laboratory shall communicate its requirements to external providers for:

a) the products and services to be provided;

b) the acceptance criteria;

c) competence, including any required qualification of personnel;

d) activities that the laboratory, or its customer, intends to perform at the external provider's premises.

ラボラトリの要求事項を，明確にし，レビュー
し，承認する．

b) 外部提供者の計価，選定，パフォーマンスの監
視及び再評価に関する基準を明確にする．

c) 外部から提供される製品及びサービスが，使用
される前又は顧客に直接提供される前に，ラボ
ラトリの設定した要求事項，又は適用可能な場
合，この規格の関連する要求事項への適合を確
実にする．

d) 外部提供者の評価，パフォーマンスの監視及び
再評価から生じた処置をとる．

6.6.3 ラボラトリは，次の事項に関して，外部提
供者に要求事項を伝達しなければならない．

a) 提供される製品及びサービス．

b) 受入基準．

c) 必要とされる要員資格を含む，力量．

d) ラボラトリ又はその顧客が外部提供者先での実
施を意図している活動．

7   Process requirements

7.1   Review of requests, tenders and contracts

7.1.1   The laboratory shall have a procedure for the review of requests, tenders and contracts. The procedure shall ensure that:

a)   the requirements are adequately defined, documented and understood;

b)   the laboratory has the capability and resources to meet the requirements;

c)   where external providers are used, the requirements of **6.6** are applied and the laboratory advises the customer of the specific laboratory activities to be performed by the external provider and gains the customer's approval;

NOTE 1   It is recognized that externally provided laboratory activities can occur when:

— the laboratory has the resources and competence to perform the activities, however, for unforeseen reasons is unable to undertake these in part or full;

7 プロセスに関する要求事項

7.1 依頼、見積仕様書及び契約のレビュー

7.1.1 ラボラトリは，依頼，見積仕様書及び契約のレビューに関する手順をもたなければならない．この手順は，次の事項を確実にしなければならない．

a) 要求事項が十分に明確化され，文書化され，理解されている．

b) ラボラトリが，要求事項を満たすための業務能力及び資源を備えている．

c) 外部提供者を利用する場合は，6.6 の要求事項が適用され，ラボラトリが顧客に対して，外部提供者によって実施される特定のラボラトリ活動に関して通知し，顧客の承認を得る．

注記 1 外部から提供されるラボラトリ活動は，次の場合に起こり得ることが認識されている．

― ラボラトリが，そのラボラトリ活動を実施する資源及び能力をもっているが，予期しなかった理由によって，その一部又は全

78   ISO/IEC 17025

— the laboratory does not have the resources
   or competence to perform the activities.

d) the appropriate methods or procedures are selected and are capable of meeting the customers' requirements.

NOTE 2   For internal or routine customers, reviews of requests, tenders and contracts can be
performed in a simplified way.

7.1.2 The laboratory shall inform the customer
when the method requested by the customer is considered to be inappropriate or out of date.

7.1.3 When the customer requests a statement
of conformity to a specification or standard for the
test or calibration (e.g. pass/fail, in-tolerance/out-
of-tolerance), the specification or standard and the
decision rule shall be clearly defined. Unless inherent in the requested specification or standard,
the decision rule selected shall be communicated

7　プロセスに関する要求事項　　79

てを実行できない場合.

—— ラボラトリが, そのラボラトリ
活動を実施する資源又は能力を
もっていない場合.

d)　適切な方法又は手順が選択され, 顧客の要求事
項を満たすことができる.

注記 2　内部の顧客又は定期の顧客に対して
は, 依頼, 見積仕様書及び契約のレビ
ューは簡素化された方法で実施するこ
とができる.

7.1.2　ラボラトリは, 顧客の依頼した方法が不適
切又は旧式であると考えられる場合, 顧客にその旨
を通知しなければならない.

7.1.3　顧客が, 試験又は校正に関して, 仕様又は
規格への適合性の表明（例えば, 合格／不合格, 許
容の範囲内／範囲外）を要請する場合は, その仕様
又は規格及び判定ルールを明確にしなければならな
い. 要請された仕様又は規格に当該取決めが内在す
る場合を除き, 選択した判定ルールを顧客に伝達し
合意を得なければならない.

80 ISO/IEC 17025

to, and agreed with, the customer.

NOTE   For further guidance on statements of conformity, see ISO/IEC Guide 98-4.

7.1.4   Any differences between the request or tender and the contract shall be resolved before laboratory activities commence. Each contract shall be acceptable both to the laboratory and the customer. Deviations requested by the customer shall not impact the integrity of the laboratory or the validity of the results.

7.1.5   The customer shall be informed of any deviation from the contract.

7.1.6   If a contract is amended after work has commenced, the contract review shall be repeated and any amendments shall be communicated to all affected personnel.

7.1.7   The laboratory shall cooperate with customers or their representatives in clarifying the customer's request and in monitoring the laboratory's

7 プロセスに関する要求事項 81

注記 適合性の表明に関する更なる手引について，ISO/IEC Guide 98-4 を参照．

7.1.4 依頼又は見積仕様書と契約との間での何らかの相違は，ラボラトリ活動が開始される前に解決しなければならない．個々の契約は，ラボラトリ及び顧客の双方にとって受入れ可能でなければならない．顧客から要請された逸脱が，ラボラトリの誠実さ（integrity）又は結果の妥当性に影響を及ぼしてはならない．

7.1.5 契約からのいかなる逸脱をも，顧客に知らせなければならない．

7.1.6 業務開始後に契約が変更される場合は，契約のレビューを繰り返さなければならない．また，全ての変更を，影響を受ける全ての要員に伝達しなければならない．

7.1.7 ラボラトリは，顧客の依頼の明確化，及び実施される業務に関連したラボラトリのパフォーマンスの監視に関して，顧客又は顧客の代理人と協力

82   ISO/IEC 17025

performance in relation to the work performed.

NOTE   Such cooperation can include:

a)   providing reasonable access to relevant areas
     of the laboratory to witness customer-specific
     laboratory activities;

b)   preparation, packaging, and dispatch of items
     needed by the customer for verification pur-
     poses.

7.1.8 Records of reviews, including any signifi-
cant changes, shall be retained. Records shall also
be retained of pertinent discussions with a customer relating to the customer's requirements or
the results of the laboratory activities.

7.2   Selection, verification and validation of methods

7.2.1   Selection and verification of methods

7.2.1.1 The laboratory shall use appropriate
methods and procedures for all laboratory activities and, where appropriate, for evaluation of the
measurement uncertainty as well as statistical

しなければならない.

注記 このような協力には，次の事項が含まれ得る.

a) 顧客固有のラボラトリ活動に立ち会うために，ラボラトリの関連する区域に正当に立ち入れるようにする.

b) 検証の目的で顧客が必要とする品目の，準備，こん（梱）包及び発送.

7.1.8 重要な変更を含め，レビューの記録を保持しなければならない．顧客の要求事項，又はラボラトリ活動の結果に関して顧客と交わした，関連する議論の記録も保持しなければならない．

7.2 方法の選定，検証及び妥当性確認

7.2.1 方法の選定及び検証

7.2.1.1 ラボラトリは，全てのラボラトリ活動に関して適切な方法及び手順を用いなければならず，また，適切な場合，測定不確かさの評価及びデータ分析のための統計的手法に関しても同様である．

84 ISO/IEC 17025

techniques for analysis of data.

NOTE "Method" as used in this document can be considered synonymous with the term "measurement procedure" as defined in ISO/IEC Guide 99.

7.2.1.2 All methods, procedures and supporting documentation, such as instructions, standards, manuals and reference data relevant to the laboratory activities, shall be kept up to date and shall be made readily available to personnel (see **8.3**).

7.2.1.3 The laboratory shall ensure that it uses the latest valid version of a method unless it is not appropriate or possible to do so. When necessary, the application of the method shall be supplemented with additional details to ensure consistent application.

NOTE International, regional or national standards or other recognized specifications that contain sufficient and concise information on how to perform laboratory activities do not need to be supplemented or rewritten as internal procedures

注記　この規格で使用される"方法"は，ISO/IEC Guide 99 において定義される"測定手順"と同義と考えられる．

7.2.1.2　全ての方法，手順，並びにラボラトリ活動に関連する指示書，規格，マニュアル及び参照データなどの支援文書は，最新の状態で維持し，要員がいつでも利用できるようにしなければならない（8.3 参照）．

7.2.1.3　ラボラトリは，有効な最新版の方法を用いることが不適切又は不可能でない限り，それを確実にしなければならない．必要な場合には，矛盾のない適用を確実にするため，詳細事項の追加によって方法の適用を補足しなければならない．

注記　ラボラトリ活動の実施方法について，国際規格，地域規格若しくは国家規格又は十分で簡潔な情報を含むその他の広く認められている仕様書が，そのままラボラトリの実施要員が使用できるように書か

if these standards are written in a way that they can be used by the operating personnel in a laboratory. It can be necessary to provide additional documentation for optional steps in the method or additional details.

7.2.1.4 When the customer does not specify the method to be used, the laboratory shall select an appropriate method and inform the customer of the method chosen. Methods published either in international, regional or national standards, or by reputable technical organizations, or in relevant scientific texts or journals, or as specified by the manufacturer of the equipment, are recommended. Laboratory-developed or modified methods can also be used.

7.2.1.5 The laboratory shall verify that it can properly perform methods before introducing them by ensuring that it can achieve the required performance. Records of the verification shall be retained. If the method is revised by the issuing body, verification shall be repeated to the extent necessary.

れている場合には，内部手順書として補足したり，書き直したりする必要はない．その方法の中での操作の選択又は詳細な補足のために，追加の文書を用意する必要があり得る．

7.2.1.4 顧客が，使用する方法を指定しない場合，ラボラトリは適切な方法を選定し，選定した方法を顧客に通知しなければならない．国際規格，地域規格若しくは国家規格のいずれかにおいて公表された方法，定評ある技術機関が公表した方法，関連する科学文献若しくは定期刊行物において公表された方法，又は設備の製造業者が指定する方法が推奨される．ラボラトリが開発又は修正した方法も用いることができる．

7.2.1.5 ラボラトリは，必要なパフォーマンスを達成できることを確実にすることによって，選定した方法を導入する前にその方法を適切に実施できることを検証しなければならない．検証の記録を保持しなければならない．その方法がそれを発行する機関によって改訂される場合，ラボラトリは，必要な程度まで検証を繰り返さなければならない．

ISO/IEC 17025

7.2.1.6 When method development is required, this shall be a planned activity and shall be assigned to competent personnel equipped with adequate resources. As method development proceeds, periodic review shall be carried out to confirm that the needs of the customer are still being fulfilled. Any modifications to the development plan shall be approved and authorized.

7.2.1.7 Deviations from methods for all laboratory activities shall occur only if the deviation has been documented, technically justified, authorized, and accepted by the customer.

NOTE   Customer acceptance of deviations can be agreed in advance in the contract.

7.2.2   Validation of methods

7.2.2.1 The laboratory shall validate non-standard methods, laboratory-developed methods and standard methods used outside their intended scope or otherwise modified. The validation shall

7 プロセスに関する要求事項　　89

7.2.1.6 方法の開発が必要な場合，これは計画的な活動でなければならず，十分な資質を備えた，力量をもつ要員に割り当てなければならない．方法の開発の進行につれて，顧客のニーズが依然として満たされていろことを確認するため，定期的な見直しを行わなければならない．開発計画の変更は，承認され，許可されなければならない．

7.2.1.7 全てのラボラトリ活動に関する方法からの逸脱は，その逸脱があらかじめ文書化され，技術的に正しいと証明され，正式に許可され，かつ，顧客によって受け入れられている場合に限らなければならない．

　　注記　逸脱に対する顧客の受入れは，契約書において事前に合意されていることもあり得る．

7.2.2　方法の妥当性確認

7.2.2.1 ラボラトリは，規格外の方法，ラボラトリが開発した方法，及び規格に規定された方法であって意図された適用範囲外で使用するもの又はその他の変更がなされたものについて，妥当性確認を行

90 ISO/IEC 17025

be as extensive as is necessary to meet the needs of the given application or field of application.

NOTE 1 Validation can include procedures for sampling, handling and transportation of test or calibration items.

NOTE 2 The techniques used for method validation can be one of, or a combination of, the following:

a) calibration or evaluation of bias and precision using reference standards or reference materials;

b) systematic assessment of the factors influencing the result;

c) testing method robustness through variation of controlled parameters, such as incubator temperature, volume dispensed;

d) comparison of results achieved with other validated methods;

e) interlaboratory comparisons;

f) evaluation of measurement uncertainty of the results based on an understanding of the theoretical principles of the method and practical

7 プロセスに関する要求事項　　91

わなければならない．妥当性確認は，特定の適用対象又は適用分野のニーズを満たすために必要な程度まで幅広く行わなければならない．

注記1　妥当性確認には，試験又は校正品目の
　　　　サンプリング，取扱い及び輸送の手順
　　　　が含まれ得る．

注記2　方法の妥当性確認に用いる手法は，次
　　　　の事項のうちの一つ又はそれらの組合
　　　　せであり得る．

　　　a)　参照標準又は標準物質を用いた，
　　　　　校正又は偏り及び精度の評価．

　　　b)　結果に影響する要因の系統的な評
　　　　　価．

　　　c)　培養器の温度，分注量などの管理
　　　　　されたパラメータの変化を通じ
　　　　　た，方法の頑健性の試験．

　　　d)　妥当性が確認された他の方法で得
　　　　　られた結果との比較．

　　　e)　試験所間比較．

　　　f)　方法の原理の理解及びサンプリン
　　　　　グ又は試験方法のパフォーマンス
　　　　　の実際の経験に基づいた，結果の

92   ISO/IEC 17025

experience of the performance of the sampling
or test method.

7.2.2.2 When changes are made to a validated
method, the influence of such changes shall be determined and where they are found to affect the
original validation, a new method validation shall
be performed.

7.2.2.3 The performance characteristics of validated methods, as assessed for the intended use,
shall be relevant to the customers' needs and consistent with specified requirements.

NOTE   Performance characteristics can include,
but are not limited to, measurement range, accuracy, measurement uncertainty of the results,
limit of detection, limit of quantification, selectiv-
ity of the method, linearity, repeatability or reproducibility, robustness against external influences
or cross-sensitivity against interference from the
matrix of the sample or test object, and bias.

7.2.2.4 The laboratory shall retain the following

7　プロセスに関する要求事項　　　93
測定不確かさの評価

7.2.2.2　妥当性が確認された方法を変更する場合は，そのような変更の影響を確定しなければならず，それらが元の妥当性確認に影響を与えることが判明した場合，新たに方法の妥当性確認を行わなければならない．

7.2.2.3　妥当性が確認された方法のパフォーマンス特性は，意図する用途に対する評価において顧客のニーズに適し，規定された要求事項に整合していなければならない．

注記　パフォーマンス特性の例には，測定範囲，精確さ，結果の測定不確かさ，検出限界，定量限界，方法の選択性，直線性，繰返し性又は再現性，外部影響に対する頑健性，又は試料若しくは試験対象のマトリックスからの干渉に対する共相関感度，及び偏りが含まれ得るが，これらに限定されない．

7.2.2.4　ラボラトリは，次の妥当性確認の記録を

94   ISO/IEC 17025

records of validation:

a)   the validation procedure used;

b)   specification of the requirements;

c)   determination of the performance characteristics of the method;

d)   results obtained;

e)   a statement on the validity of the method, detailing its fitness for the intended use.

7.3 Sampling

7.3.1 The laboratory shall have a sampling plan and method when it carries out sampling of substances, materials or products for subsequent testing or calibration. The sampling method shall address the factors to be controlled to ensure the validity of subsequent testing or calibration results. The sampling plan and method shall be available at the site where sampling is undertaken. Sampling plans shall, whenever reasonable, be based on appropriate statistical methods.

7.3.2 The sampling method shall describe:

a)   the selection of samples or sites;

保持しなければならない.

a) 使用した妥当性確認の手順.

b) 要求事項の詳述.

c) 方法のパフォーマンス特性の確定.

d) 得られた結果.

e) 意図した用途に対する方法の適切性を詳述した,方法の妥当性に関する表明.

7.3 サンプリング

7.3.1 ラボラトリは,後の試験又は校正のための物質,材料又は製品のサンプリングを実施する場合,サンプリングの計画及び方法をもたなければならない.サンプリング方法は,後の試験又は校正結果の妥当性を確実にするために管理すべき要因を考慮しなければならない.サンプリングの計画及び方法は,サンプリングが行われる場所で利用できなければならない.サンプリング計画は,合理的である限り,適切な統計的方法に基づかなければならない.

7.3.2 サンプリング方法は,次の事項を記述しなければならない.

a) サンプル又はサンプリング場所の選択

96   ISO/IEC 17025

b)   the sampling plan;

c)   the preparation and treatment of sample(s) from a substance, material or product to yield the required item for subsequent testing or calibration.

NOTE   When received into the laboratory, further handling can be required as specified in **7.4**.

7.3.3   The laboratory shall retain records of sampling data that forms part of the testing or calibration that is undertaken. These records shall include, where relevant:

a)   reference to the sampling method used;

b)   date and time of sampling;

c)   data to identify and describe the sample (e.g. number, amount, name);

d)   identification of the personnel performing sampling;

e)   identification of the equipment used;

f)   environmental or transport conditions;

g)   diagrams or other equivalent means to identify the sampling location, when appropriate;

7 プロセスに関する要求事項　　97

b) サンプリング計画

c) 後の試験又は校正のために必要な品目を得るための、物質、材料又は製品からのサンプルの準備及び処理

> **注記** ラボラトリに受領された際に，**7.4** に規定された更なる取扱いが必要とされる場合がある．

7.3.3 ラボラトリは，請け負った試験・校正の一部を構成する該当サンプリングデータの記録を保持しなければならない．これらの記録には，該当する場合，次の事項を含めなければならない．

a) 用いたサンプリング手順の参照．

b) サンプリングの日付及び時刻．

c) 試料を特定し記述するためのデータ（例えば，数，量，名称）．

d) サンプリングを実施した要員の識別．

e) 使用された設備の識別．

f) 環境条件又は輸送条件．

g) 適切な場合，サンプリング場所を特定するための図面又はその他の同等な手段．

98 ISO/IEC 17025

h) deviations, additions to or exclusions from the sampling method and sampling plan.

7.4 Handling of test or calibration items

7.4.1 The laboratory shall have a procedure for the transportation, receipt, handling, protection, storage, retention, and disposal or return of test or calibration items, including all provisions necessary to protect the integrity of the test or calibration item, and to protect the interests of the laboratory and the customer. Precautions shall be taken to avoid deterioration, contamination, loss or damage to the item during handling, transporting, storing/waiting, and preparation for testing or calibration. Handling instructions provided with the item shall be followed.

7.4.2 The laboratory shall have a system for the unambiguous identification of test or calibration items. The identification shall be retained while the item is under the responsibility of the laboratory. The system shall ensure that items will not be confused physically or when referred to in records or other documents. The system shall, if ap-

7　プロセスに関する要求事項　　99

h) サンプリング方法及びサンプリング計画からの
逸脱，追加又は除外．

7.4　試験・校正品目の取扱い

7.4.1 ラボラトリは，試験・校正品目の完全性並
びにラボラトリ及び顧客の利益を保護するために必
要な全ての規定を含め，試験・校正品目の輸送，受
領，取扱い，保護，保管，保留及び処分又は返却の
ための手順をもたなければならない．ラボラトリ
は，試験又は校正のための取扱い，輸送，保管／待
機及び準備の間に品目が劣化，汚染，損失又は損傷
を受けることを防止するための予防策をとらなけれ
ばならない．試験・校正品目に添えられた取扱いの
指示に従わなければならない．

7.4.2 ラボラトリは，試験・校正品目の明確な識
別のためのシステムをもたなければならない．この
識別は，当該品目がラボラトリの責任下にある間，
保持されなければならない．識別システムは，品目
の物理的な混同又は記録若しくはその他の文書で引
用する際の混同が起こらないことを確実にしなけれ
ばならない．識別システムは，適切ならば品目又は

100   ISO/IEC 17025

propriate, accommodate a sub-division of an item or groups of items and the transfer of items.

7.4.3 Upon receipt of the test or calibration item, deviations from specified conditions shall be recorded. When there is doubt about the suitability of an item for test or calibration, or when an item does not conform to the description provided, the laboratory shall consult the customer for further instructions before proceeding and shall record the results of this consultation. When the customer requires the item to be tested or calibrated acknowledging a deviation from specified conditions, the laboratory shall include a disclaimer in the report indicating which results may be affected by the deviation.

7.4.4 When items need to be stored or conditioned under specified environmental conditions, these conditions shall be maintained, monitored and recorded.

7.5 Technical records

7.5.1 The laboratory shall ensure that technical

品目のグループの小分け及び品目の移送に対応しなければならない.

7.4.3 試験・校正品目を受領した際,規定された状態からの逸脱を記録しなければならない.品目の試験・校正に対する適性に何らかの疑義がある場合,又は品目が添えられた記述に適合しない場合,ラボラトリは,業務を進める前に更なる指示を求めて顧客に相談し,この相談の結果を記録しなければならない.顧客が,規定された状態からの逸脱を認めながらその品目の試験又は校正を要求する場合,ラボラトリは,その逸脱によってどの結果が影響を受けるおそれがあるのかを示した免責条項を報告書に含めなければならない.

7.4.4 規定された環境条件下で品目を保管又は調整する必要がある場合は,これらの条件を維持し,監視し,記録しなければならない.

7.5 技術的記録

7.5.1 ラボラトリは,個々のラボラトリ活動の技

102   ISO/IEC 17025

records for each laboratory activity contain the results, report and sufficient information to facilitate, if possible, identification of factors affecting the measurement result and its associated measurement uncertainty and enable the repetition of the laboratory activity under conditions as close as possible to the original. The technical records shall include the date and the identity of personnel responsible for each laboratory activity and for checking data and results. Original observations, data and calculations shall be recorded at the time they are made and shall be identifiable with the specific task.

7.5.2   The laboratory shall ensure that amendments to technical records can be tracked to previous versions or to original observations. Both the original and amended data and files shall be retained, including the date of alteration, an indication of the altered aspects and the personnel responsible for the alterations.

7.6   Evaluation of measurement uncertainty

7.6.1   Laboratories shall identify the contribu-

術的記録には，結果，報告並びに可能であれば測定結果及び付随する測定不確かさに影響を与える要因の特定を容易にし，元の条件にできるだけ近い条件でラボラトリ活動の反復を可能とする十分な情報が含まれることを確実にしなければならない．その技術的記録には，日付並びに個々のラボラトリ活動及びデータ・結果の確認に責任をもつ要員の識別を含めなければならない．観測原本，データ及び計算は，それらが作成される時点において記録され，特定の業務において識別可能でなければならない．

7.5.2　ラボラトリは，技術的記録の変更について，以前の版又は観測原本に遡って追跡できることを確実にしなければならない．変更の日付，変更点の表示及び変更に責任をもつ要員を含め，元のデータ及び変更されたデータ並びにそれらのファイルの両方を保持しなければならない．

7.6　測定不確かさの評価

7.6.1　ラボラトリは，測定不確かさへの寄与成分

104   ISO/IEC 17025

tions to measurement uncertainty. When evaluating measurement uncertainty, all contributions that are of significance, including those arising from sampling, shall be taken into account using appropriate methods of analysis.

7.6.2 A laboratory performing calibrations, including of its own equipment, shall evaluate the measurement uncertainty for all calibrations.

7.6.3 A laboratory performing testing shall evaluate measurement uncertainty. Where the test method precludes rigorous evaluation of measurement uncertainty, an estimation shall be made based on an understanding of the theoretical principles or practical experience of the performance of the method.

NOTE 1   In those cases where a well-recognized test method specifies limits to the values of the major sources of measurement uncertainty and specifies the form of presentation of the calculated results, the laboratory is considered to have satisfied **7.6.3** by following the test method and report-

7 プロセスに関する要求事項 105

を特定しなければならない．測定不確かさを評価する際，サンプリングから生じるものを含み，重大な全ての寄与成分を，適切な分析方法を用いて考慮しなければならない．

7.6.2 校正を実施するラボラトリは，所有する設備を含め，全ての校正に関する測定不確かさを評価しなければならない．

7.6.3 試験を実施するラボラトリは，測定不確かさを評価しなければならない．試験方法によって，厳密な測定不確かさの評価ができない場合，原理の理解又は試験方法の実施に関する実際の経験に基づいて推定しなければならない．

注記 **1** 広く認められた試験方法が，測定不確かさの主な要因の値に限界を定め，計算結果の表現形式を規定している場合には，ラボラトリは，試験方法及び報告方法の指示に従うことによって，**7.6.3** を満足しているとみなされる．

ing instructions.

NOTE 2   For a particular method where the measurement uncertainty of the results has been established and verified, there is no need to evaluate measurement uncertainty for each result if the laboratory can demonstrate that the identified critical influencing factors are under control.

NOTE 3   For further information, see ISO/IEC Guide 98-3, ISO 21748 and the ISO 5725 series.

7.7   Ensuring the validity of results

7.7.1   The laboratory shall have a procedure for monitoring the validity of results. The resulting data shall be recorded in such a way that trends are detectable and, where practicable, statistical techniques shall be applied to review the results. This monitoring shall be planned and reviewed and shall include, where appropriate, but not be limited to:

a)   use of reference materials or quality control materials;

b)   use of alternative instrumentation that has been calibrated to provide traceable results;

7 プロセスに関する要求事項　　107

注記 2　結果の測定不確かさが確立され，検証されている特定の方法に関して特定された重大な影響因子が制御されていることをラボラトリが実証できる場合，個々の結果について測定不確かさを評価する必要はない．

注記 3　さらに詳しい情報については，**ISO/IEC Guide 98-3**，**JIS Z 8404-1** 及び **JIS Z 8402** 規格群を参照．

7.7　結果の妥当性の確保

7.7.1　ラボラトリは，結果の妥当性を監視するための手順をもたなければならない．結果として得られるデータは，傾向が検出できるような方法で記録し，実行可能な場合，結果のレビューに統計的手法を適用しなければならない．この監視は，計画し，見直さなければならない．また，適切な場合，次の事項を含めなければならないが，これらに限定されない．

a)　標準物質又は品質管理用物質の使用

b)　トレーサブルな結果を得るために校正された代替の計測機器の使用

108   ISO/IEC 17025

c)   functional check(s) of measuring and testing equipment;

d)   use of check or working standards with control charts, where applicable;

e)   intermediate checks on measuring equipment;

f)   replicate tests or calibrations using the same or different methods;

g)   retesting or recalibration of retained items;

h)   correlation of results for different characteristics of an item;

i)   review of reported results;

j)   intralaboratory comparisons;

k)   testing of blind sample(s).

7.7.2 The laboratory shall monitor its performance by comparison with results of other laboratories, where available and appropriate. This monitoring shall be planned and reviewed and shall include, but not be limited to, either or both of the following:

a)   participation in proficiency testing;

      NOTE   ISO/IEC 17043 contains additional information on proficiency tests and proficiency testing providers. Proficiency testing pro-

7 プロセスに関する要求事項　　109

c) 測定設備及び試験設備の機能チェック

d) 適用可能な場合，チェック標準又は実用標準の管理図を伴う使用

e) 測定設備の中間チェック

f) 同じ方法又は異なる方法を用いた試験又は校正の反復

g) 保留された品目の再試験又は再校正

h) 一つの品目の異なる特性に関する結果の相関

i) 報告された結果のレビュー

j) 試験所内比較

k) ブラインドサンプルの試験

7.7.2　ラボラトリは，利用可能で適切な場合，他のラボラトリの結果との比較によって，そのパフォーマンスを監視しなければならない．この監視は，計画し，見直さなければならない．また，次のいずれか，又は両方を含まなければならないが，これらに限定されない．

a) 技能試験への参加

　　　注記　**JIS Q 17043** は，技能試験及び技能試験提供者に関する追加情報を含んでいる．**JIS Q 17043** の要求事項を満

viders that meet the requirements of ISO/IEC 17043 are considered to be competent.

b) participation in interlaboratory comparisons other than proficiency testing.

7.7.3 Data from monitoring activities shall be analysed, used to control and, if applicable, improve the laboratory's activities. If the results of the analysis of data from monitoring activities are found to be outside pre-defined criteria, appropriate action shall be taken to prevent incorrect results from being reported.

7.8 Reporting of results
7.8.1 General
7.8.1.1 The results shall be reviewed and authorized prior to release.

7.8.1.2 The results shall be provided accurately, clearly, unambiguously and objectively, usually in a report (e.g. a test report or a calibration certificate or report of sampling), and shall include all the information agreed with the customer and necessary for the interpretation of the results and

7　プロセスに関する要求事項　　111

たす技能試験提供者は，能力があると
みなされる．

h)　技能試験以外の試験所間比較への参加

7.7.3　ラボラトリは，監視活動で得られたデータ
を分析し，ラボラトリ活動の管理に使用し，適用可
能であれば，改善に使用しなければならない．監視
活動で得られたデータの分析結果が，事前に規定し
た処置基準を外れることが判明した場合は，不正確
な結果が報告されることを防止するため，適切な処
置を講じなければならない．

7.8　結果の報告
7.8.1　一般
7.8.1.1　結果は，開示する前に，レビューされ，
承認されなければならない．

7.8.1.2　結果は，通常，報告書（例えば，試験報
告書，校正証明書又はサンプリング報告書）の形
で，正確に，明瞭に，曖昧でなく，客観的に提供さ
れなければならない．また，結果には，顧客と合意
し，かつ，結果の解釈に必要な全ての情報及び用い
た方法が要求する全ての情報を含めなければならな

112 ISO/IEC 17025

all information required by the method used. All issued reports shall be retained as technical records.

NOTE 1   For the purposes of this document, test reports and calibration certificates are sometimes referred to as test certificates and calibration reports, respectively.

NOTE 2   Reports can be issued as hard copies or by electronic means, provided that the requirements of this document are met.

7.8.1.3   When agreed with the customer, the results may be reported in a simplified way. Any information listed in **7.8.2** to **7.8.7** that is not reported to the customer shall be readily available.

7.8.2   Common requirements for reports (test, calibration or sampling)

7.8.2.1   Each report shall include at least the following information, unless the laboratory has valid reasons for not doing so, thereby minimizing

い．発行された全ての報告書は，技術的記録として保持しなければならない．

注記1 この規格の目的において，試験報告書及び校正証明書は，それぞれ試験証明書及び校正報告書と呼ばれることがある．

注記2 この規格の要求事項が満たされている限り，報告書はハードコピー又は電子的手段によって発行することができる．

7.8.1.3 顧客との合意がある場合には，簡略化した方法で結果を報告してもよい．**7.8.2 〜 7.8.7** に規定されているが，顧客に報告されなかったいかなる情報も，すぐに利用できるようにしておかなければならない．

7.8.2 報告書（試験，校正又はサンプリング）に関する共通の要求事項

7.8.2.1 個々の報告書は，少なくとも次の情報を含まなければならない．ただし，ラボラトリが正当な除外の理由をもち，それによって誤解又は誤用の

114   ISO/IEC 17025

any possibility of misunderstanding or misuse:

a) a title (e.g. "Test Report", "Calibration Certificate" or "Report of Sampling");

b) the name and address of the laboratory;

c) the location of performance of the laboratory activities, including when performed at a customer facility or at sites away from the laboratory's permanent facilities, or in associated temporary or mobile facilities;

d) unique identification that all its components are recognized as a portion of a complete report and a clear identification of the end;

e) the name and contact information of the customer;

f) identification of the method used;

g) a description, unambiguous identification, and, when necessary, the condition of the item;

h) the date of receipt of the test or calibration item(s), and the date of sampling, where this is critical to the validity and application of the results;

i) the date(s) of performance of the laboratory activity;

j) the date of issue of the report;

可能性が最小化される場合はこの限りでない.

a) タイトル（例えば，"試験報告書"，"校正証明書" 又は "サンプリング報告書"）

h) ラボラトリの名称及び住所

c) 顧客の施設若しくはラボラトリの恒久的施設から離れた場所，又は関連する一時施設若しくは移動施設で実施された場合を含め，ラボラトリ活動が実施された場所

d) 全ての構成要素が完全な報告書の一部であることが分かる固有の識別，及び報告書の終わりを示す明瞭な識別

e) 顧客の名称及び連絡先情報

f) 用いた方法の識別

g) 品目の記述，明確な識別，及び必要な場合，品目の状態

h) 結果の妥当性及び適用に重大な意味をもつ場合は，試験・校正品目の受領日，及びサンプリングの実施日

i) ラボラトリ活動の実施日（期間）

j) 報告書の発行日

116 ISO/IEC 17025

k) reference to the sampling plan and sampling method used by the laboratory or other bodies where these are relevant to the validity or application of the results;

l) a statement to the effect that the results relate only to the items tested, calibrated or sampled;

m) the results with, where appropriate, the units of measurement;

n) additions to, deviations, or exclusions from the method;

o) identification of the person(s) authorizing the report;

p) clear identification when results are from external providers.

NOTE   Including a statement specifying that the report shall not be reproduced except in full without approval of the laboratory can provide assurance that parts of a report are not taken out of context.

7.8.2.2 The laboratory shall be responsible for all the information provided in the report, except

7 プロセスに関する要求事項　　117

k)　サンプリング計画及びサンプリング方法が結果
　　の妥当性又は適用に関連する場合には，ラボラ
　　トリ又はその他の機関が用いたサンプリング計
　　画及びサンプリング方法の参照

l)　結果が，その試験，校正又はサンプリングされ
　　た品目だけに関するものであるという旨の表明

m)　結果．適切な場合，測定単位を伴う．

n)　方法への追加又は方法からの逸脱若しくは除外

o)　報告書の承認権限者の識別

p)　結果が外部提供者から出されたものである場合
　　は，明確な識別

　　注記　ラボラトリの承認なく報告書の一部分だ
　　　　　けを複製してはならないことを規定する
　　　　　表明を含めることによって，報告書の一
　　　　　部が前後関係から切り離されないことを
　　　　　保証することができる．

7.8.2.2　ラボラトリは，その情報が顧客から提供
されたものである場合を除き，報告書に記載された

when information is provided by the customer. Data provided by a customer shall be clearly identified. In addition, a disclaimer shall be put on the report when the information is supplied by the customer and can affect the validity of results. Where the laboratory has not been responsible for the sampling stage (e.g. the sample has been provided by the customer), it shall state in the report that the results apply to the sample as received.

7.8.3 Specific requirements for test reports

7.8.3.1 In addition to the requirements listed in **7.8.2**, test reports shall, where necessary for the interpretation of the test results, include the following:

a) information on specific test conditions, such as environmental conditions;

b) where relevant, a statement of conformity with requirements or specifications (see **7.8.6**);

c) where applicable, the measurement uncertainty presented in the same unit as that of the measurand or in a term relative to the measurand (e.g. percent) when:

— it is relevant to the validity or application

全ての情報について責任をもたなければならない．顧客によって提供されたデータは，明確に識別されなければならない．さらに，その情報が顧客から提供されたもので，結果の妥当性に影響する可能性がある場合には，免責条項を報告書に記載しなければならない．ラボラトリがサンプリング段階に責任をもたない場合（例えば，試料が顧客から提供された場合）には，結果は受領した試料に適用される旨を報告書に記載しなければならない．

7.8.3 試験報告書に関する特定要求事項

7.8.3.1 **7.8.2** の要求事項に加え，試験結果の解釈に必要な場合，試験報告書は次の事項を含まなければならない．

a) 特定の試験条件に関する情報，例えば，環境条件

b) 該当する場合，要求事項又は仕様に対する適合性の表明（**7.8.6** 参照）

c) 適用可能な場合であって，次のいずれかの条件を満たす場合には，測定対象量と同じ単位で表示された，又は測定対象量に対する相対値（例えば，パーセント）で表示された測定不確かさ

— 測定不確かさが，試験結果の妥当性又は適用

of the test results;

— a customer's instruction so requires, or

— the measurement uncertainty affects conformity to a specification limit;

d) where appropriate, opinions and interpretations (see **7.8.7**);

e) additional information that may be required by specific methods, authorities, customers or groups of customers.

7.8.3.2 Where the laboratory is responsible for the sampling activity, test reports shall meet the requirements listed in **7.8.5** where necessary for the interpretation of test results.

7.8.4 Specific requirements for calibration certificates

7.8.4.1 In addition to the requirements listed in **7.8.2**, calibration certificates shall include the following:

a) the measurement uncertainty of the measurement result presented in the same unit as that of the measurand or in a term relative to the

に関連している.

— 顧客の指示が,測定不確かさを要求している.

— 測定不確かさが,仕様の限界への適合性に影響を与える.

d) 適切な場合,意見及び解釈(**7.8.7** 参照)

e) 特定の方法,規制当局,顧客又は顧客のグループによって要求されることがある追加の情報

7.8.3.2 ラボラトリがサンプリング活動に責任をもつ場合,試験結果の解釈に必要であれば,試験報告書は,**7.8.5** の要求事項を満たさなければならない.

7.8.4 校正証明書に関する特定要求事項

7.8.4.1 **7.8.2** の要求事項に加え,校正証明書は,次の事項を含まなければならない.

a) 測定対象量と同じ単位で表示された,又は測定対象量に対する相対値(例えば,パーセント)で表示された測定結果についての測定不確かさ

122   ISO/IEC 17025

measurand (e.g. percent);

NOTE   According to ISO/IEC Guide 99, a measurement result is generally expressed as a single measured quantity value including unit of measurement and a measurement uncertainty.

b)   the conditions (e.g. environmental) under which the calibrations were made that have an inf luence on the measurement results;

c)   a statement identifying how the measurements are metrologically traceable (see **Annex A**);

d)   the results before and after any adjustment or repair, if available;

e)   where relevant, a statement of conformity with requirements or specifications (see **7.8.6**);

f)   where appropriate, opinions and interpretations (see **7.8.7**).

7.8.4.2   Where the laboratory is responsible for the sampling activity, calibration certificates shall meet the requirements listed in **7.8.5** where necessary for the interpretation of calibration results.

7 プロセスに関する要求事項　　123

　　注記　**ISO/IEC Guide 99** によれば，測定
　　　　結果は一般に，測定の単位及び測定不
　　　　確かさを含む，単一の測定された量の
　　　　値で表される．

b) 測定結果に影響をもつ，校正が実施された際の
　　条件（例えば，環境条件）

c) 測定値がどのように計量トレーサビリティをも
　　つのかを明確化した表明（**附属書 A を参照**）

d) 利用可能な場合，調整又は修理の前後の結果

e) 該当する場合，要求事項又は仕様への適合性の
　　表明（**7.8.6 参照**）
f) 適切な場合，意見及び解釈（**7.8.7 参照**）

7.8.4.2　ラボラトリがサンプリング活動に責任を
もつ場合，校正結果の解釈に必要であれば，校正証
明書は，**7.8.5** の要求事項を満たさなければならな
い．

7.8.4.3 A calibration certificate or calibration label shall not contain any recommendation on the calibration interval, except where this has been agreed with the customer.

7.8.5 Reporting sampling – specific requirements

Where the laboratory is responsible for the sampling activity, in addition to the requirements listed in **7.8.2**, reports shall include the following, where necessary for the interpretation of results:

a)   the date of sampling;

b)   unique identification of the item or material sampled (including the name of the manufacturer, the model or type of designation and serial numbers, as appropriate);

c)   the location of sampling, including any diagrams, sketches or photographs;

d)   a reference to the sampling plan and sampling method;

e)   details of any environmental conditions during sampling that affect the interpretation of the results;

f)   information required to evaluate measure-

7.8.4.3 顧客との合意がある場合を除き，校正証明書又は校正ラベルのいずれも，校正周期に関する推奨を含んではならない．

7.8.5 サンプリングの報告－特定要求事項

ラボラトリがサンプリング活動に責任をもつ場合，**7.8.2** に列挙する要求事項に加え，結果の解釈に必要な場合には，報告書は次の事項を含まなければならない．

a) サンプリングの日付

b) サンプリングされた品目又は材料の固有の識別（適切な場合，製造業者の名称，指定されたモデル又は型式，及びシリアル番号を含む．）

c) 図面，スケッチ又は写真を含む，サンプリングの場所

d) サンプリングの計画及び方法の参照

e) 結果の解釈に影響する，サンプリング中の環境条件の詳細

f) 後の試験又は校正の測定不確かさを評価するた

126   ISO/IEC 17025

ment uncertainty for subsequent testing or calibration.

7.8.6   Reporting statements of conformity

7.8.6.1   When a statement of conformity to a specification or standard is provided, the laboratory shall document the decision rule employed, taking into account the level of risk (such as false accept and false reject and statistical assumptions) associated with the decision rule employed, and apply the decision rule.

NOTE   Where the decision rule is prescribed by the customer, regulations or normative documents, a further consideration of the level of risk is not necessary.

7.8.6.2   The laboratory shall report on the statement of conformity, such that the statement clearly identifies:

a)   to which results the statement of conformity applies;

b)   which specifications, standards or parts thereof are met or not met;

めに必要な情報

7.8.6 適合性の表明の報告

7.8.6.1 ラボラトリは，仕様又は規格への適合性を表明する場合，採用した判定ルールに付随する，（誤判定による合格及び誤判定による不合格，並びに統計的仮定などの）リスクのレベルを考慮に入れた上で採用した判定ルールを文書化し，それを適用しなければならない．

注記 その判定ルールが，顧客，規制又は規範文書によって規定されている場合，リスクのレベルの更なる検討は不要である．

7.8.6.2 ラボラトリは，次の事項を明示して，適合性の表明に関する報告を行わなければならない．

a) どの結果に対して適合性の表明が適用されるのか．

b) どの仕様，規格又はそれらの一部に適合又は不適合なのか．

c) the decision rule applied (unless it is inherent in the requested specification or standard).

NOTE   For further information, see ISO/IEC Guide 98-4.

7.8.7   Reporting opinions and interpretations

7.8.7.1   When opinions and interpretations are expressed, the laboratory shall ensure that only personnel authorized for the expression of opinions and interpretations release the respective statement. The laboratory shall document the basis upon which the opinions and interpretations have been made.

NOTE   It is important to distinguish opinions and interpretations from statements of inspections and product certifications as intended in ISO/IEC 17020 and ISO/IEC 17065, and from statements of conformity as referred to in **7.8.6**.

7.8.7.2   The opinions and interpretations expressed in reports shall be based on the results obtained from the tested or calibrated item and shall

7 プロセスに関する要求事項 129

c) 適用された判定ルール（要求された仕様又は規格に既に含まれている場合を除く.）.

> 注記 さらに詳しい情報については，ISO/IEC Guide 98-4 を参照.

7.8.7 意見及び解釈の報告

7.8.7.1 ラボラトリは，意見及び解釈を表明する場合，それらを表明する権限を与えられた要員だけがそれぞれの表明を提示することを確実にしなければならない.ラボラトリは，意見及び解釈が形成された根拠を文書化しなければならない.

> 注記 意見及び解釈は，**JIS Q 17020** 及び **JIS Q 17065** が意図している検査及び製品認証の表明，並びに **7.8.6** の適合性の表明と区別することが重要である.

7.8.7.2 報告書に表明する意見及び解釈は，試験又は校正した品目から得られた結果に基づかなければならず，意見及び解釈である旨を明示しなければ

130   ISO/IEC 17025

be clearly identified as such.

7.8.7.3 When opinions and interpretations are directly communicated by dialogue with the customer, a record of the dialogue shall be retained.

7.8.8 Amendments to reports

7.8.8.1 When an issued report needs to be changed, amended or re-issued, any change of information shall be clearly identified and, where appropriate, the reason for the change included in the report.

7.8.8.2 Amendments to a report after issue shall be made only in the form of a further document, or data transfer, which includes the statement "Amendment to Report, serial number... [or as otherwise identified]", or an equivalent form of wording.

Such amendments shall meet all the requirements of this document.

7.8.8.3 When it is necessary to issue a complete new report, this shall be uniquely identified and

ならない.

7.8.7.3 意見及び解釈が顧客との対話で直接伝達される場合，その対話の記録を保持しなければならない.

7.8.8 報告書の修正

7.8.8.1 発行済みの報告書を変更，修正又は再発行する必要がある場合は，いかなる情報の変更も明確に識別し，適切な場合，変更の理由を報告書に含めなければならない.

7.8.8.2 発行後の報告書の修正は，"報告書，シリアル番号 ...（又は他の識別）の修正"という表明若しくは同等の文言を含めた，追加文書又はデータ転送という形態だけによって行わなければならない.

そのような修正は，この規格の全ての要求事項を満たさなければならない.

7.8.8.3 完全な新規の報告書を発行することが必要な場合には，この新規の報告書に固有の識別を与

shall contain a reference to the original that it replaces.

7.9 Complaints

7.9.1 The laboratory shall have a documented process to receive, evaluate and make decisions on complaints.

7.9.2 A description of the handling process for complaints shall be available to any interested party on request. Upon receipt of a complaint, the laboratory shall confirm whether the complaint relates to laboratory activities that it is responsible for and, if so, shall deal with it. The laboratory shall be responsible for all decisions at all levels of the handling process for complaints.

7.9.3 The process for handling complaints shall include at least the following elements and methods:

a)   description of the process for receiving, validating, investigating the complaint, and deciding what actions are to be taken in response to it;

え，それが置き換わる元の報告書の引用を含めなければならない．

7.9 苦情

7.9.1 ラボラトリは，苦情を受領し，評価し，決定を下すための文書化したプロセスをもたなければならない．

7.9.2 苦情処理プロセスの記述は，いかなる利害関係者にも，要請に応じて入手可能にしなければならない．苦情を受領した時点で，ラボラトリは，その苦情が，自らが責任をもつラボラトリ活動に関係するかどうかを確認し，関係があればその苦情を処理しなければならない．ラボラトリは，苦情処理プロセスの全ての階層において，全ての決定について責任をもたなければならない．

7.9.3 苦情処理プロセスは，少なくとも次の要素及び方法を含まなければならない．

a) 苦情を受領し，妥当性を確認し，調査を行い，それに対応してとるべき処置を決定するためのプロセスを記述する．

134   ISO/IEC 17025

b)   tracking and recording complaints, including actions undertaken to resolve them;

c)   ensuring that any appropriate action is taken.

7.9.4   The laboratory receiving the complaint shall be responsible for gathering and verifying all necessary information to validate the complaint.

7.9.5   Whenever possible, the laboratory shall acknowledge receipt of the complaint, and provide the complainant with progress reports and the outcome.

7.9.6   The outcomes to be communicated to the complainant shall be made by, or reviewed and approved by, individual(s) not involved in the original laboratory activities in question.

NOTE   This can be performed by external personnel.

7.9.7   Whenever possible, the laboratory shall give formal notice of the end of the complaint handling to the complainant.

7 プロセスに関する要求事項　　135

h)　苦情を解決するためにとられる処置を含め，苦情を追跡し，記録する．

c)　適切な処置がとられることを確実にする．

7.9.4　苦情を受領するラボラトリは，その苦情の妥当性を確認するために必要な全ての情報の収集及び検証に責任をもたなければならない．

7.9.5　ラボラトリは，可能な場合には，苦情申立者に対して苦情の受領を通知し，進捗状況及び結果を提示しなければならない．

7.9.6　苦情申立者に伝達される結果は，問題となっている元のラボラトリ活動に関与していなかった者が作成するか，又はレビューし承認しなければならない．

　　注記　これは，外部の要員によって実施することができる．

7.9.7　ラボラトリは，可能な場合には，苦情処理の終了を苦情申立者に対して正式に通知しなければならない．

7.10 Nonconforming work

7.10.1 The laboratory shall have a procedure that shall be implemented when any aspect of its laboratory activities or results of this work do not conform to its own procedures or the agreed requirements of the customer (e.g. equipment or environmental conditions are out of specified limits, results of monitoring fail to meet specified criteria). The procedure shall ensure that:

a) the responsibilities and authorities for the management of nonconforming work are defined;

b) actions (including halting or repeating of work and withholding of reports, as necessary) are based upon the risk levels established by the laboratory;

c) an evaluation is made of the significance of the nonconforming work, including an impact analysis on previous results;

d) a decision is taken on the acceptability of the nonconforming work;

e) where necessary, the customer is notified and work is recalled;

f) the responsibility for authorizing the resump-

7.10 不適合業務

7.10.1　ラボラトリは，そのラボラトリ活動の何らかの業務の側面，又はその結果が，ラボラトリの手順又は顧客との間で合意された要求事項に適合しない場合（例えば，設備又は環境条件が規定の限界を外れている場合，監視の結果が規定の基準を満たさない場合）に実施しなければならない手順をもたなければならない．この手順は，次の事項を確実にしなければならない．

a)　不適合業務の管理に関する責任及び権限を定める．

b)　処置（必要に応じて，業務を停止する又は繰り返すこと，及び報告書を保留することを含む．）を，ラボラトリの設定したリスクレベルに基づいて定める．

c)　以前の結果に関する影響分析を含め，不適合業務の重大さを評価する．

d)　不適合業務の容認の可否を決定する．

e)　必要な場合，顧客に通知して業務結果を回収する．

f)　業務の再開を承認する責任を定める．

138   ISO/IEC 17025

tion of work is defined.

7.10.2   The laboratory shall retain records of non-conforming work and actions as specified in **7.10.1**, bullets b) to f).

7.10.3   Where the evaluation indicates that the nonconforming work could recur, or that there is doubt about the conformity of the laboratory's operations with its own management system, the laboratory shall implement corrective action.

7.11 Control of data and information management

7.11.1   The laboratory shall have access to the data and information needed to perform laboratory activities.

7.11.2   The laboratory information management system(s) used for the collection, processing, recording, reporting, storage or retrieval of data shall be validated for functionality, including the proper functioning of interfaces within the laboratory information management system(s) by the

7.10.2 ラボラトリは，不適合業務及び **7.10.1** の
b) 〜 f) に規定する処置の記録を保持しないればな
らない．

7.10.3 ラボラトリは，評価によって，不適合業務
が再発し得ること又はラボラトリ自身のマネジメン
トシステムに対する運営の適合性に疑いがあること
が示された場合には，是正処置を実施しなければな
らない．

7.11 データの管理及び情報マネジメント

7.11.1 ラボラトリは，ラボラトリ活動を行うため
に必要なデータ及び情報を利用できなければならな
い．

7.11.2 データの収集，処理，記録，報告，保管又
は検索に使用されるラボラトリ情報マネジメントシ
ステムは，導入の前に，ラボラトリによって，ラボ
ラトリ情報マネジメントシステム内のインタフェー
スが適正に機能していることを含め，機能性の妥当
性を確認しなければならない．ラボラトリ情報マネ

140 ISO/IEC 17025

laboratory before introduction. Whenever there are any changes, including laboratory software configuration or modifications to commercial off-the-shelf software, they shall be authorized, documented and validated before implementation.

NOTE 1   In this document "laboratory information management system(s)" includes the management of data and information contained in both computerized and non-computerized systems. Some of the requirements can be more applicable to computerized systems than to non-computerized systems.

NOTE 2   Commercial off-the-shelf software in general use within its designed application range can be considered to be sufficiently validated.

7.11.3   The laboratory information management system(s) shall:

a)   be protected from unauthorized access;

b)   be safeguarded against tampering and loss;

c)   be operated in an environment that complies

7 プロセスに関する要求事項　　141

ジメントシステムは，ラボラトリによるソフトウェアの設定変更又は市販の既製ソフトウェアの変更を含め，変更が行われる場合には，使用前に承認し，文書化し，妥当性を確認しなければならない．

注記1　この規格において，"ラボラトリ情報マネジメントシステム"には，電子化されたシステム及び電子化されていないシステムの両方に含まれるデータ並びに情報の管理が含まれる．要求事項によっては，電子化されていないシステムより電子化されているシステムに適用しやすくなる．

注記2　一般的に使用されている市販の既製ソフトウェアは，設計上の適用範囲において十分に妥当性が確認されているとみなすことができる．

7.11.3　ラボラトリ情報マネジメントシステムは，次の事項を満たさなければならない．

a)　無許可のアクセスから保護されている．

b)　不正な書き換え及び損失から防護されている．

c)　提供者若しくはラボラトリの仕様に適合する環

142   ISO/IEC 17025

with provider or laboratory specifications or, in the case of non-computerized systems, provides conditions which safeguard the accuracy of manual recording and transcription;

d)   be maintained in a manner that ensures the integrity of the data and information;

e)   include recording system failures and the appropriate immediate and corrective actions.

7.11.4   When a laboratory information management system is managed and maintained off-site or through an external provider, the laboratory shall ensure that the provider or operator of the system complies with all applicable requirements of this document.

7.11.5   The laboratory shall ensure that instructions, manuals and reference data relevant to the laboratory information management system(s) are made readily available to personnel.

7.11.6   Calculations and data transfers shall be checked in an appropriate and systematic manner.

境の中で運用されているか，又は電子化されていないシステムの場合は，手書きの記録及び転記の正確さを確保する条件を備える環境の中で運用されている．

d) データ及び情報の完全性を確実にする方法で維持されている．

e) システム障害及びそれに対する適切な応急処置及び是正処置を記録することを含む．

7.11.4 ラボラトリ情報マネジメントシステムが，異なる場所（off-site）で管理及び保守されているか，又は外部提供者を通じて管理及び保守されている場合，ラボラトリは，システムの提供者又は操作者が，この規格の適用される全ての要求事項に適合することを確実にしなければならない．

7.11.5 ラボラトリは，ラボラトリ情報マネジメントシステムに関連する指示書，マニュアル及び参照データを要員がいつでも利用できることを確実にしなければならない．

7.11.6 計算及びデータ転送は，適切かつ系統的な方法でチェックを行わなければならない．

8 Management system requirements

8.1 Options

8.1.1 General

The laboratory shall establish, document, implement and maintain a management system that is capable of supporting and demonstrating the consistent achievement of the requirements of this document and assuring the quality of the laboratory results. In addition to meeting the requirements of **Clauses 4** to **7**, the laboratory shall implement a management system in accordance with Option A or Option B.

NOTE   See **Annex B** for more information.

8.1.2 Option A

As a minimum, the management system of the laboratory shall address the following:

— management system documentation (see **8.2**);

— control of management system documents (see **8.3**);

— control of records (see **8.4**);

— actions to address risks and opportunities (see **8.5**);

8 マネジメントシステムに関する要求事項

8.1 選択肢

8.1.1 一般

ラボラトリは，この規格の要求事項の一貫した達成を支援し，実証するとともに，試験・校正結果の品質を保証することを可能にするマネジメントシステムを構築し，文書化し，実施し，維持しなければならない．この規格の箇条4〜箇条7の要求事項に適合することに加え，ラボラトリは，選択肢A又は選択肢Bに基づくマネジメントシステムを実施しなければならない．

注記　詳しくは，附属書Bを参照．

8.1.2 選択肢A

ラボラトリのマネジメントシステムは，少なくとも次の事項に取り組まなければならない．

— マネジメントシステムの文書化（8.2参照）
— マネジメントシステム文書の管理（8.3参照）

— 記録の管理（8.4参照）
— リスク及び機会への取組み（8.5参照）

146   ISO/IEC 17025

— improvement (see **8.6**);

— corrective actions (see **8.7**);

— internal audits (see **8.8**);

— management reviews (see **8.9**).

8.1.3 Option B

A laboratory that has established and maintains a management system, in accordance with the requirements of ISO 9001, and that is capable of supporting and demonstrating the consistent fulfilment of the requirements of **Clauses 4** to **7**, also fulfils at least the intent of the management system requirements specified in **8.2** to **8.9**.

8.2 Management system documentation (Option A)

8.2.1 Laboratory management shall establish, document, and maintain policies and objectives for the fulfilment of the purposes of this document and shall ensure that the policies and objectives are acknowledged and implemented at all levels of the laboratory organization.

8.2.2 The policies and objectives shall address

改善（**8.6** 参照）

— 是正処置（**8.7** 参照）

— 内部監査（**8.8** 参照）

— マネジメントレビュー（**8.9** 参照）

8.1.3 選択肢 B

JIS Q 9001 の要求事項に従ってマネジメントシステムを確立し，維持しており，この規格の箇条 4 〜箇条 7 の要求事項を一貫して満たすことを裏付け，実証することが可能なラボラトリは，少なくとも **8.2** 〜 **8.9** に規定するマネジメントシステム要求事項の意図をも満たしている．

8.2 マネジメントシステムの文書化（選択肢 A）

8.2.1 ラボラトリマネジメントは，この規格の目的を果たすための方針及び目標を，確立し，文書化し，維持し，ラボラトリの組織の全ての階層で，この方針及び目標が周知され，実施されることを確実にしなければならない．

8.2.2 この方針及び目標は，ラボラトリの能力，

148   ISO/IEC 17025

the competence, impartiality and consistent operation of the laboratory.

8.2.3   Laboratory management shall provide evidence of commitment to the development and implementation of the management system and to
continually improving its effectiveness.

8.2.4   All documentation, processes, systems, records, related to the fulfilment of the requirements
of this document shall be included in, referenced
from, or linked to the management system.

8.2.5   All personnel involved in laboratory activities shall have access to the parts of the management system documentation and related information that are applicable to their responsibilities.

8.3   Control of management system documents (Option A)

8.3.1   The laboratory shall control the documents
(internal and external) that relate to the fulfilment of this document.

公平性及び一貫性のある運営を取り上げていなければならない．

8.2.3 ラボラトリマネジメントは，マネジメントシステムの開発及び実施，並びにマネジメントシステムの有効性の継続的改善に対するコミットメントの証拠を提示しなければならない．

8.2.4 この規格の要求事項を満たすことに関係する全ての文書，プロセス，システム，記録をマネジメントシステムに含めるか，マネジメントシステムから引用するか，又はマネジメントシステムに関連付けなければならない．

8.2.5 ラボラトリ活動に関与する全ての要員は，それらの要員の職責に適用されるマネジメントシステム文書及び関連情報の該当部分を利用できなければならない．

8.3 マネジメントシステム文書の管理（選択肢 A）

8.3.1 ラボラトリは，この規格を満たすことに関係する（内部及び外部の）文書を管理しなければならない．

150   ISO/IEC 17025

NOTE   In this context, "documents" can be policy statements, procedures, specifications, manufacturer's instructions, calibration tables, charts, text books, posters, notices, memoranda, drawings, plans, etc. These can be on various media, such as hard copy or digital.

8.3.2   The laboratory shall ensure that:

a)   documents are approved for adequacy prior to issue by authorized personnel;

b)   documents are periodically reviewed, and updated as necessary;

c)   changes and the current revision status of documents are identified;

d)   relevant versions of applicable documents are available at points of use and, where necessary, their distribution is controlled;

e)   documents are uniquely identified;

f)   the unintended use of obsolete documents is prevented, and suitable identification is applied to them if they are retained for any purpose.

注記　ここでいう "文書" とは，方針表明文，手順書，仕様書，製造業者の指示書，校正価表，チャート，教科書，ポスター，通知，覚書，図面，図解などであり得る．それらは，ハードコピーか，又はデジタル形式のような様々な媒体で作成できる．

8.3.2　ラボラトリは，次の事項を確実にしなければならない．

a)　文書の発行に先立って，権限をもった要員がその文書の妥当性について承認を与える．

b)　文書を定期的に見直し，必要に応じて更新する．

c)　文書の変更及び最新の改訂の状況が識別される．

d)　適用される文書の適切な版が使用に際して入手でき，必要に応じてそれらの文書の配布が管理される．

e)　文書に固有の識別を付す．

f)　廃止文書の意図しない使用を防止する．目的を問わず，廃止文書を保持する場合は，それらに適切な識別を付す．

8.4 Control of records (Option A)

8.4.1 The laboratory shall establish and retain legible records to demonstrate fulfilment of the requirements in this document.

8.4.2 The laboratory shall implement the controls needed for the identification, storage, protection, back-up, archive, retrieval, retention time, and disposal of its records. The laboratory shall retain records for a period consistent with its contractual obligations. Access to these records shall be consistent with the confidentiality commitments, and records shall be readily available.

NOTE   Additional requirements regarding technical records are given in **7.5**.

8.5 Actions to address risks and opportunities (Option A)

8.5.1 The laboratory shall consider the risks and opportunities associated with the laboratory activities in order to:

a)   give assurance that the management system achieves its intended results;

8.4 記録の管理 （選択肢 A）

8.4.1 ラボラトリは，この規格の要求事項を満たすことを実証するための読みやすい記録を確立し，保持しなければならない．

8.4.2 ラボラトリは，記録の識別，保管，保護，バックアップ，アーカイブ，検索，保持期間及び廃棄のために必要な管理を実施しなければならない．ラボラトリは，契約上の義務に準じた期間にわたって記録を保持しなければならない．これらの記録へのアクセスは，機密保持のコミットメントに準じなければならない．また，記録は直ちに利用できなければならない．

> **注記** 技術的記録に関する追加的な要求事項は，**7.5** に記載されている．

8.5 リスク及び機会への取組み （選択肢 A）

8.5.1 ラボラトリは，次の事項を目的として，ラボラトリ活動に付随するリスク及び機会を考慮しなければならない．

a) マネジメントシステムが，その意図した結果を達成できるという確信を与える．

b)   enhance opportunities to achieve the purpose
     and objectives of the laboratory;

c)   prevent, or reduce, undesired impacts and po-
     tential failures in the laboratory activities;

d)   achieve improvement.

8.5.2   The laboratory shall plan:

a)   actions to address these risks and opportuni-
     ties;

b)   how to:

     —   integrate and implement these actions
         into its management system;

     —   evaluate the effectiveness of these ac-
         tions.

NOTE   Although this document specifies that
the laboratory plans actions to address risks,
there is no requirement for formal methods for
risk management or a documented risk management process. Laboratories can decide whether or
not to develop a more extensive risk management
methodology than is required by this document,
e.g. through the application of other guidance or

8 マネジメントシステムに関する要求事項 155

b) ラボラトリの目的及び目標を達成する機会を広げる.

c) ラボラトリ活動における望ましくない影響及び潜在的障害を防止又は低減する.

d) 改善を達成する.

8.5.2 ラボラトリは,次の事項を計画しなければならない.

a) これらのリスク及び機会への取組み.

b) 次の事項を行う方法.
— これらの取組みのマネジメントシステムへの統合及び実施.
— これらの取組みの有効性の評価.

注記 この規格は,ラボラトリのリスクへの取組みの計画について規定するが,リスクマネジメントの正式な方法又は文書化されたリスクマネジメントプロセスの要求事項は規定していない.ラボラトリは,例えば,他の手引又は規格の適用を通じて,この規格によって要求されるリスクマネジメント手法よりも広範な手法を開

156 ISO/IEC 17025

standards.

8.5.3 Actions taken to address risks and opportunities shall be proportional to the potential impact on the validity of laboratory results.

NOTE 1 Options to address risks can include identifying and avoiding threats, taking risk in order to pursue an opportunity, eliminating the risk source, changing the likelihood or consequences, sharing the risk, or retaining risk by informed decision.

NOTE 2 Opportunities can lead to expanding the scope of the laboratory activities, addressing new customers, using new technology and other possibilities to address customer needs.

8.6 Improvement (Option A)

8.6.1 The laboratory shall identify and select opportunities for improvement and implement any necessary actions.

NOTE Opportunities for improvement can be identified through the review of the operational

8 マネジメントシステムに関する要求事項 157

発するか否かを決定できる.

8.5.3 リスク及び機会への取組みは,ラボラトリが出す結果の妥当性に与える潜在的影響に釣り合ったものでなければならない.

注記1 リスクへの取組みの選択肢には,脅威の特定及び回避,機会を追求するためのリスク負担,リスク源の除去,可能性若しくは結果の変更,リスクの共有,又は情報に基づく決定によるリスク保持が含まれ得る.

注記2 機会は,ラボラトリ活動の範囲拡大,新たな顧客への取組み,新技術の使用及び顧客のニーズに取り組むその他の可能性につながり得る.

8.6 改善(選択肢A)

8.6.1 ラボラトリは,改善の機会を特定し,選択して,必要な処置を実施しなければならない.

注記 改善の機会は,業務手順のレビュー,方針の使用,全体の目標,監査結果,是正

158 ISO/IEC 17025

procedures, the use of the policies, overall objectives, audit results, corrective actions, management review, suggestions from personnel, risk assessment, analysis of data, and proficiency testing results.

8.6.2 The laboratory shall seek feedback, both positive and negative, from its customers. The feedback shall be analysed and used to improve the management system, laboratory activities and customer service.

NOTE   Examples of the types of feedback include customer satisfaction surveys, communication records and review of reports with customers.

8.7   Corrective actions (Option A)

8.7.1   When a nonconformity occurs, the laboratory shall:

a)   react to the nonconformity and, as applicable:

— take action to control and correct it;

8 マネジメントシステムに関する要求事項　159

　　処置，マネジメントレビュー，要員から
　　の提案，リスクアセスメント，データの
　　分析，技能試験の結果を通じて特定する
　　ことができる．

8.6.2　ラボラトリは，顧客からの肯定的なフィー
ドバック及び否定的なフィードバックの両方を求め
なければならない．マネジメントシステム，ラボラ
トリ活動及び顧客へのサービスの改善のためにフィ
ードバックを分析し，利用しなければならない．

　　注記　フィードバックの種類の例には，顧客満
　　　　足の調査，コミュニケーションの記録及
　　　　び顧客と共同での報告書のレビューが含
　　　　まれる．

8.7　是正処置（選択肢 A）

8.7.1　不適合が発生した場合，ラボラトリは，次
の事項を行わなければならない．

a)　その不適合に対処し，該当する場合には，必
　　ず，次の事項を行う．

　　—　その不適合を管理し，修正するための処置を
　　　　とる．

160　　　　　　ISO/IEC 17025

— address the consequences;

b) evaluate the need for action to eliminate the cause(s) of the nonconformity, in order that it does not recur or occur elsewhere, by:

— reviewing and analysing the nonconformity;

— determining the causes of the nonconformity;

— determining if similar nonconformities exist, or could potentially occur;

c) implement any action needed;

d) review the effectiveness of any corrective action taken;

e) update risks and opportunities determined during planning, if necessary;

f) make changes to the management system, if necessary.

8.7.2 Corrective actions shall be appropriate to the effects of the nonconformities encountered.

8.7.3 The laboratory shall retain records as evidence of:

8 マネジメントシステムに関する要求事項　161

― その不適合の結果に対処する.

b) その不適合が再発又は他のところで発生しないようにするため，次の事項によって，その不適合の原因を除去するための処置をとる必要性を評価する.

― 不適合をレビューし，分析する.

― その不適合の原因を明確にする.

― 類似の不適合の有無，又はそれらが発生する可能性を明確にする.

c) 必要な処置を実施する.

d) とった全ての是正処置の有効性をレビューする.

e) 必要な場合には，計画の過程で明確になったリスク及び機会を更新する.

f) 必要な場合には，マネジメントシステムの変更を行う.

8.7.2 是正処置は，検出された不適合のもつ影響に応じたものでなければならない.

8.7.3 ラボラトリは，次の事項の証拠として記録を保持しなければならない.

162 ISO/IEC 17025

a) the nature of the nonconformities, cause(s) and any subsequent actions taken;

b) the results of any corrective action.

8.8 Internal audits (Option A)

8.8.1 The laboratory shall conduct internal audits at planned intervals to provide information on whether the management system:

a) conforms to:

— the laboratory's own requirements for its management system, including the laboratory activities;

— the requirements of this document;

b) is effectively implemented and maintained.

8.8.2 The laboratory shall:

a) plan, establish, implement and maintain an audit programme including the frequency, methods, responsibilities, planning requirements and reporting, which shall take into consideration the importance of the laboratory activities concerned, changes affecting the

8 マネジメントシステムに関する要求事項 163

a) 不適合の性質，原因及びそれに対してとったあらゆる処置

b) 是正処置の結果

8.8 内部監査（選択肢 A）

8.8.1 ラボラトリは，マネジメントシステムが次の状況にあるか否かに関する情報を提供するために，あらかじめ定めた間隔で内部監査を実施しなければならない．

a) 次の事項に適合している．

— ラボラトリ活動を含めた，ラボラトリ自体のマネジメントシステムに関する要求事項

— この規格の要求事項

b) 有効に実施され，維持されている．

8.8.2 ラボラトリは，次の事項を行わなければならない．

a) 頻度，方法，責任，要求事項の立案，及び報告を含む，監査プログラムを計画し，確立し，実施し，維持する．監査プログラムは，関連するラボラトリ活動の重要性，ラボラトリに影響を及ぼす変更及び前回までの監査の結果を考慮に入れなければならない．

164   ISO/IEC 17025

laboratory, and the results of previous audits;

b)   define the audit criteria and scope for each audit;

c)   ensure that the results of the audits are reported to relevant management;

d)   implement appropriate correction and corrective actions without undue delay;

e)   retain records as evidence of the implementation of the audit programme and the audit results.

NOTE   ISO 19011 provides guidance for internal audits.

8.9   Management reviews (Option A)

8.9.1   The laboratory management shall review its management system at planned intervals, in order to ensure its continuing suitability, adequacy and effectiveness, including the stated policies and objectives related to the fulfilment of this document.

8.9.2   The inputs to management review shall be recorded and shall include information related to

b) 各監査について，監査基準及び監査範囲を定める．

c) 監査の結果を関連する管理要員に報告することを確実にする．

d) 遅滞なく，適切な修正及び是正処置を実施する．

e) 監査プログラムの実施及び監査結果の証拠として，記録を保持する．

　　注記　**JIS Q 19011** は，内部監査に関する指針を示している．

8.9　マネジメントレビュー（選択肢 A）

8.9.1　ラボラトリマネジメントは，マネジメントシステムが引き続き，適切，妥当かつ有効であることを確実にするために，この規格を満たすことに関係する明示された方針及び目標を含め，あらかじめ定めた間隔でマネジメントシステムをレビューしなければならない．

8.9.2　マネジメントレビューへのインプットは，記録しなければならない．また，マネジメントレビ

166 ISO/IEC 17025

the following:

a) changes in internal and external issues that are relevant to the laboratory;

b) fulfilment of objectives;

c) suitability of policies and procedures;

d) status of actions from previous management reviews;

e) outcome of recent internal audits;

f) corrective actions;

g) assessments by external bodies;

h) changes in the volume and type of the work or in the range of laboratory activities;

i) customer and personnel feedback;

j) complaints;

k) effectiveness of any implemented improvements;

l) adequacy of resources;

m) results of risk identification;

n) outcomes of the assurance of the validity of results; and

o) other relevant factors, such as monitoring activities and training.

8 マネジメントシステムに関する要求事項　167

ューへのインプットには，次の事項に関係する情報を含めなければならない．

a) ラボラトリに関連する，内部及び外部の課題の変化

b) 目標の達成

c) 方針及び手順の適切さ

d) 前回までのマネジメントレビューの結果とった処置の状況

e) 最近の内部監査の結果

f) 是正処置

g) 外部機関による評価

h) 業務の量及び種類の変化，又はラボラトリ活動の範囲の変更

i) 顧客及び要員からのフィードバック

j) 苦情

k) 実施された改善の有効性

l) 資源の適切性

m) リスク特定の結果

n) 結果の妥当性の保証の成果

o) 監視活動及び教育訓練などのその他の関連因子

8.9.3 The outputs from the management review shall record all decisions and actions related to at least:

a)   the effectiveness of the management system and its processes;

b)   improvement of the laboratory activities related to the fulfilment of the requirements of this document;

c)   provision of required resources;

d)   any need for change.

8　マネジメントシステムに関する要求事項　169

8.9.3　マネジメントレビューからのアウトプットは，少なくとも次の事項に関係する全ての決定及び処置を記録しなければならない

a)　マネジメントシステム及びそのプロセスの有効性

b)　この規格の要求事項を満たすことに関係するラボラトリ活動の改善

c)　必要とされる資源の提供

d)　あらゆる変更の必要性

Annex A

(informative)

Metrological traceability

A.1 General

This annex provides additional information on metrological traceability, which is an important concept to ensure comparability of measurement results both nationally and internationally.

A.2 Establishing metrological traceability

A.2.1 Metrological traceability is established by considering, and then ensuring, the following:

a) the specification of the measurand (quantity to be measured);

b) a documented unbroken chain of calibrations going back to stated and appropriate references (appropriate references include national or international standards, and intrinsic standards);

c) that measurement uncertainty for each step in the traceability chain is evaluated according to agreed methods;

附属書 A

(参考)

計量トレーサビリティ

A.1　一般

計量トレーサビリティは，国内的にも国際的にも，測定結果の比較可能性を確実にするために重要な概念であり，この附属書では計量トレーサビリティに関する追加的情報を記載する．

A.2　計量トレーサビリティの確立

A.2.1　計量トレーサビリティは，次の事項を検討し，確実にすることによって確立される．

a)　測定対象量の詳述．

b)　定められた，適切な計量参照（適切な計量参照には，国家標準又は国際標準，及び固有標準が含まれる．）まで遡る，文書化された，切れ目のない校正の連鎖．

c)　トレーサビリティの連鎖の各段階の測定不確かさは，合意された方法に従って評価される．

172   ISO/IEC 17025

d)   that each step of the chain is performed in accordance with appropriate methods, with tho measurement results and with associated, recorded measurement uncertainties;

e)   that the laboratories performing one or more steps in the chain supply evidence for their technical competence.

A.2.2   The systematic measurement error (sometimes called "bias") of the calibrated equipment is taken into account to disseminate metrological traceability to measurement results in the laboratory. There are several mechanisms available to take into account the systematic measurement errors in the dissemination of measurement metrological traceability.

A.2.3   Measurement standards that have reported information from a competent laboratory that includes only a statement of conformity to a specification (omitting the measurement results and associated uncertainties) are sometimes used to disseminate metrological traceability. This approach, in which the specification limits are imported as

d) 連鎖の各段階は，適切な方法で実施されており，測定結果及び付随し記録された測定不確かさが伴う．

e) 連鎖の一段階以上を実施するラボラトリは，技術的能力に関する証拠を提示する．

A.2.2 ラボラトリにおける測定結果に計量トレーサビリティを与えるために，校正された設備の系統測定誤差（"偏り"と呼ばれることがある．）を考慮に入れる．計量トレーサビリティの供給において系統測定誤差を考慮に入れる幾つかの方法がある．

A.2.3 ある仕様に対する適合性の表明（測定結果と付随する不確かさを省略）だけを含む，能力のあるラボラトリから報告された情報をもった測定標準が，計量トレーサビリティを与えるものとして用いられることがある．仕様限界が不確かさ要因として組み入れられているこのアプローチは，次の事項に依存する．

the source of uncertainty, is dependent upon:

— the use of an appropriate decision rule to establish conformity;

— the specification limits subsequently being treated in a technically appropriate way in the uncertainty budget.

The technical basis for this approach is that the declared conformance to a specification defines a range of measurement values, within which the true value is expected to lie, at a specified level of confidence, which considers both any bias from the true value, as well as the measurement uncertainty.

EXAMPLE   The use of OIML R 111 class weights to calibrate a balance.

A.3   Demonstrating metrological traceability

A.3.1   Laboratories are responsible for establishing metrological traceability in accordance with this document. Calibration results from laboratories conforming to this document provide metrological traceability. Certified values of certified

— 適合性を確立するための適切な判定ルールの使用

— 仕様限界が不確かさバジェットにおいて，技術的に適切なアプローチで取り扱われる事項

このアプローチの技術的根拠は，仕様に対して宣言された適合性が測定値の範囲を明確にしており，そこでは，ある指定された信頼の水準において，真値がその範囲内にあると期待され，真値からの偏りも，測定不確かさも考慮していることである．

例 はかりを校正するために用いられる **OIML R 111** に等級が規定された分銅の使用

A.3 計量トレーサビリティの実証

A.3.1 ラボラトリは，この規格に基づいて計量トレーサビリティを確立する責任をもつ．この規格に適合するラボラトリの校正結果は，計量トレーサビリティを提供する．**JIS Q 17034** に適合する標準物質生産者からの認証標準物質の認証値は，計量ト

176   ISO/IEC 17025

reference materials from reference material producers conforming to ISO 17034 provide metrological traceability. There are various ways to demonstrate conformity with this document: third party recognition (such as an accreditation body), external assessment by customers or self-assessment. Internationally accepted paths include, but are not limited to, the following.

a)   Calibration and measurement capabilities provided by national metrology institutes and designated institutes that have been subject to suitable peer-review processes. Such peer-review is conducted under the CIPM MRA (International Committee for Weights and Measures Mutual Recognition Arrangement). Services covered by the CIPM MRA can be viewed in Appendix C of the BIPM KCDB (International Bureau of Weights and Measures Key Comparison Database) which details the range and measurement uncertainty for each listed service.

b)   Calibration and measurement capabilities that have been accredited by an accreditation body subject to the ILAC (International Laboratory

附属書 A（参考）

レーサビリティを提供する．第一者による承認（認定機関等），顧客による外部評価，自己評価といった，この規格への適合を実証する様々な方法がある．国際的に受け入れられる方法には次のものが含まれるが，これらに限定されない．

a) CIPM MRA（国際度量衡委員会相互承認取決め）等の国際的取決めの下でピアレビュー・プロセスを経た校正測定能力．CIPM MRA の対象となるサービスは，BIPM（国際度量衡局）KCDB（基幹比較データベース）の附属書 C で見ることができる．このデータベースには，一覧に記載された各サービスの範囲及び不確かさが含まれる．

b) ILAC（国際試験所認定協力機構）取決め，又は ILAC が承認した地域取決めに基づく認定機関によって認定を受けた校正測定能力は，計量

Accreditation Cooperation) Arrangement or to Regional Arrangements recognized by ILAC have demonstrated metrological traceability. Scopes of accredited laboratories are publicly available from their respective accreditation bodies.

A.3.2 The Joint BIPM, OIML (International Organization of Legal Metrology), ILAC and ISO Declaration on Metrological Traceability provides specific guidance when there is a need to demonstrate international acceptability of the metrological traceability chain.

附属書 A（参考）

トレーサビリティを実証している．認定された校正機関の範囲は，それぞれの認定機関から公に入手できる．

A.3.2 計量トレーサビリティに関する BIPM, OIML（国際法定計量機関），ILAC 及び **ISO** の共同宣言は，計量トレーサビリティの連鎖が国際的に受け入れられることを実証することが必要な場合の明確な指針を定めている．

Annex B

(informative)

Management system options

B.1 Growth in the use of management systems generally has increased the need to ensure that laboratories can operate a management system that is seen as conforming to ISO 9001, as well as to this document. As a result, this document provides two options for the requirements related to the implementation of a management system.

B.2 Option A (see **8.1.2**) lists the minimum requirements for implementation of a management system in a laboratory. Care has been taken to incorporate all those requirements of ISO 9001 that are relevant to the scope of laboratory activities that are covered by the management system. Laboratories that comply with **Clauses 4** to **7** and implement Option A of **Clause 8** will therefore also operate generally in accordance with the principles of ISO 9001.

附属書 B

(参考)

マネジメントシステムに関する選択肢

B.1 マネジメントシステムの利用が増加したことによって，ラボラトリが，この規格に加えて **JIS Q 9001** にも適合するとみなされるマネジメントシステムを運用できることを確実にする必要性が，一般的に高まった．その結果，この規格では，マネジメントシステムの実施に関係する要求事項について，二つの選択肢を示すことになった．

B.2 選択肢 A (8.1.2 参照) は，ラボラトリにおけるマネジメントシステムの実施に関する最小限の要求事項を列挙したものである．マネジメントシステムが対象とする，ラボラトリ活動の範囲に該当する **JIS Q 9001** の全ての要求事項を取り入れるよう注意が払われた．したがって，箇条 4 〜箇条 7 に適合し，かつ，箇条 8 の選択肢 A を実施するラボラトリは，一般に **JIS Q 9001** の原則にも従って運営されていることになる．

B.3 Option B (see **8.1.3**) allows laboratories to establish and maintain a management system in accordance with the requirements of ISO 9001, in a manner that supports and demonstrates the consistent fulfilment of **Clauses 4** to **7**. Laboratories that implement Option B of **Clause 8** will therefore also operate in accordance with ISO 9001. Conformity of the management system within which the laboratory operates to the requirements of ISO 9001 does not, in itself, demonstrate the competence of the laboratory to produce technically valid data and results. This is accomplished through compliance with **Clauses 4** to **7**.

B.4 Both options are intended to achieve the same result in the performance of the management system and compliance with **Clauses 4** to **7**.

NOTE   Documents, data and records are components of documented information as used in ISO 9001 and other management system standards. Control of documents is covered in **8.3**. The control of records is covered in **8.4** and **7.5**. The control of data related to the laboratory activities is covered

B.3 選択肢 B（**8.1.3** 参照）によって，この規格の箇条 4 〜 箇条 7 を一貫して満たしていることを裏付け，実証するという形をとることで，ラボラトリは，**JIS Q 9001** の要求事項に従ってマネジメントシステムを確立し，維持することが可能となる．したがって，箇条 8 の選択肢 B を実施するラボラトリは，**JIS Q 9001** にも従って運営されていることになる．ラボラトリが運営しているマネジメントシステムが **JIS Q 9001** の要求事項に適合していることは，それ自体ではそのラボラトリが技術的に有効なデータ及び結果を生成する能力をもつことを実証するわけではない．これは，この規格の箇条 4 〜箇条 7 の順守を通じて実現される．

B.4 いずれの選択肢も，マネジメントシステムのパフォーマンス及び箇条 4 〜 箇条 7 の順守において，同じ結果を達成することを意図している．

　　注記　文書，データ及び記録は，**JIS Q 9001** 及びその他のマネジメントシステム規格で用いられている，文書化された情報の構成要素である．文書の管理については，**8.3** に記載している．記録の管理については，**8.4** 及び **7.5** に記載してい

in **7.11**.

B.5 **Figure B.1** illustrates an example of a possible schematic representation of the operational processes of a laboratory, as described in **Clause 7**.

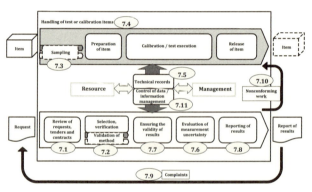

Figure B.1— **Possible schematic representation of the operational processes of a laboratory**

る．ラボラトリ活動に関係するデータの管理については，7.11 に記載している．

B.5 図 B.1 は，箇条 7 に記載しているような，ラボラトリの運用プロセスの図解の例を示す．

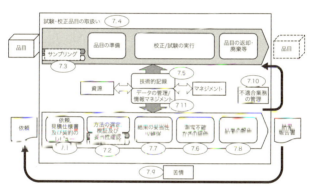

図 B.1 －ラボラトリの運用プロセスの図解

Bibliography

[1] ISO 5725-1, *Accuracy (trueness and precision) of measurement methods and results — Part 1: General principles and definitions*

[2] ISO 5725-2, *Accuracy (trueness and precision) of measurement methods and results — Part 2: Basic method for the determination of repeatability and reproducibility of a standard measurement method*

[3] ISO 5725-3, *Accuracy (trueness and precision) of measurement methods and results — Part 3: Intermediate measures of the precision of a standard measurement method*

参考文献

[1] **JIS Z 8402-1** 測定方法及び測定結果の精確さ
（真度及び精度） 第1部：一般的な原理及び定義

注記 対応国際規格：**ISO 5725-1**, Accuracy
(trueness and precision) of measurement methods and results—Part 1:
General principles and definitions

[2] **JIS Z 8402-2** 測定方法及び測定結果の精確さ
（真度及び精度）—第2部：標準測定方法の併行
精度及び再現精度を求めるための基本的方法

注記 対応国際規格：**ISO 5725-2**, Accuracy
(trueness and precision) of measurement methods and results—Part 2:
Basic method for the determination
of repeatability and reproducibility
of a standard measurement method

[3] **JIS Z 8402-3** 測定方法及び測定結果の精確さ
（真度及び精度）—第3部：標準測定方法の中間
精度

注記 対応国際規格：**ISO 5725-3**, Accuracy
(trueness and precision) of measurement methods and results—Part 3:
Intermediate measures of the pre-

[4]   ISO 5725-4, *Accuracy (trueness and precision) of measurement methods and results — Part 4: Basic methods for the determination of the trueness of a standard measurement method*

[5]   ISO 5725-6, *Accuracy (trueness and precision) of measurement methods and results — Part 6: Use in practice of accuracy values*

[6]   ISO 9000, *Quality management systems — Fundamentals and vocabulary*

[7]   ISO 9001, *Quality management systems — Re-*

cision of a standard measurement method

[4] **JIS Z 8402-4** 測定方法及び測定結果の精確さ（真度及び精度）—第4部：標準測定方法の真度を求めるための基本的方法

注記 対応国際規格：**ISO 5725-4**, Accuracy (trueness and precision) of measurement methods and results—Part 4: Basic methods for the determination of the trueness of a standard measurement method

[5] **JIS Z 8402-6** 測定方法及び測定結果の精確さ（真度及び精度）—第6部：精確さに関する値の実用的な使い方

注記 対応国際規格：**ISO 5725-6**, Accuracy (trueness and precision) of measurement methods and results—Part 6: Use in practice of accuracy values

[6] **JIS Q 9000** 品質マネジメントシステム—基本及び用語

注記 対応国際規格：**ISO 9000**, Quality management systems—Fundamentals and vocabulary

[7] **JIS Q 9001** 品質マネジメントシステム—要求

quirements

[8] ISO 10012, *Measurement management systems — Requirements for measurement processes and measuring equipment*

[9] ISO/IEC 12207, *Systems and software engineering — Software life cycle processes*

[10] ISO 15189, *Medical laboratories — Requirements for quality and competence*

[11] ISO 15194, *In vitro diagnostic medical devices — Measurement of quantities in samples of biological origin — Requirements for certified reference materials and the content of supporting documentation*

[12] ISO/IEC 17011, *Conformity assessment — Requirements for accreditation bodies accredit-*

事項

注記　対応国際規格：ISO 9001, Quality management systems — Requirements

[8]　**JIS Q 10012**　計測マネジメントシステム — 測定プロセス及び測定機器に関する要求事項

注記　対応国際規格：**ISO 10012**, Measurement management systems — Requirements for measurement processes and measuring equipment

[9]　**JIS X 0160**　ソフトウェアライフサイクルプロセス

注記 1　対応国際規格：**ISO/IEC 12207**, Systems and software engineering — Software life cycle processes

注記 2　**ISO/IEC 12207** は，**ISO/IEC/IEEE 12207**:2017 へ改訂されている．

[10]　**ISO 15189**, Medical laboratories — Requirements for quality and competence

[11]　**ISO 15194**, In vitro diagnostic medical devices — Measurement of quantities in samples of biological origin — Requirements for certified reference materials and the content of supporting documentation

[12]　**JIS Q 17011**　適合性評価 — 適合性評価機関の認定を行う機関に対する要求事項

ing conformity assessment bodies

[13] ISO/IEC 17020, *Conformity assessment — Requirements for the operation of various types of bodies performing inspection*

[14] ISO/IEC 17021-1, *Conformity assessment — Requirements for bodies providing audit and certification of management systems — Part 1: Requirements*

[15] ISO 17034, *General requirements for the competence of reference material producers*

注記 対応国際規格：**ISO/IEC 17011**, Conformity assessment—Requirements for accreditation bodies accrediting conformity assessment bodies

[13] **JIS Q 17020** 適合性評価—検査を実施する各種機関の運営に関する要求事項

注記 対応国際規格：**ISO/IEC 17020**, Conformity assessment—Requirements for the operation of various types of bodies performing inspection

[14] **JIS Q 17021-1** 適合性評価—マネジメントシステムの審査及び認証を行う機関に対する要求事項—第1部：要求事項

注記 対応国際規格：**ISO/IEC 17021-1**, Conformity assessment—Requirements for bodies providing audit and certification of management systems—Part 1: Requirements

[15] **JIS Q 17034** 標準物質生産者の能力に関する一般要求事項

注記 対応国際規格：**ISO 17034**, General requirements for the competence of reference material producers

[16] ISO/IEC 17043, *Conformity assessment — General requirements for proficiency testing*

[17] ISO/IEC 17065, *Conformity assessment — Requirements for bodies certifying products, processes and services*

[18] ISO 17511, *In vitro diagnostic medical devices — Measurement of quantities in biological samples — Metrological traceability of values assigned to calibrators and control materials*

[19] ISO 19011, *Guidelines for auditing management systems*

[20] ISO 21748, *Guidance for the use of repeatability, reproducibility and trueness estimates in*

参考文献 195

[16] **JIS Q 17043** 適合性評価―技能試験に対する一般要求事項

> **注記** 対応国際規格：ISO/IEC 17043, Conformity assessment —General requirements for proficiency testing

[17] **JIS Q 17065** 適合性評価―製品, プロセス及びサービスの認証を行う機関に対する要求事項

> **注記** 対応国際規格：ISO/IEC 17065, Conformity assessment —Requirements for bodies certifying products, processes and services

[18] **ISO 17511**, In vitro diagnostic medical devices—Measurement of quantities in biological samples—Metrological traceability of values assigned to calibrators and control materials

[19] **JIS Q 19011** マネジメントシステム監査のための指針

> **注記** 対応国際規格：ISO 19011, Guidelines for auditing management systems

[20] **JIS Z 8404-1** 測定の不確かさ―第1部：測定の不確かさの評価における併行精度, 再現精

196 ISO/IEC 17025

measurement uncertainty evaluation

[21] ISO 31000, *Risk management — Guidelines*

[22] ISO Guide 30, *Reference materials — Selected terms and definitions*

[23] ISO Guide 31, *Reference materials — Contents of certificates, labels and accompanying documentation*

[24] ISO Guide 33, *Reference materials — Good practice in using reference materials*

度及び真度の推定値の利用の指針

> 注記 対応国際規格：**ISO 21748**, Guidance for the use of repeatability, reproducibility and trueness estimates in measurement uncertainty evaluation

[21] **JIS Q 31000** リスクマネジメント―原則及び指針

> 注記 対応国際規格：**ISO 31000**, Risk management―Principles and guidelines

[22] **JIS Q 0030** 標準物質に関連して用いられる用語及び定義

> 注記 対応国際規格：**ISO Guide 30**, Reference materials―Selected terms and definitions

[23] **JIS Q 0031** 標準物質―認証書，ラベル及び附属文書の内容

> 注記 対応国際規格：**ISO Guide 31**, Reference materials―Contents of certificates, labels and accompanying documentation

[24] **JIS Q 0033** 認証標準物質の使い方

> 注記 対応国際規格：**ISO Guide 33**, Ref-

[25] ISO Guide 35, *Reference materials — Guidance for characterization and assessment of homogeneity and stability*

[26] ISO Guide 80, *Guidance for the in-house preparation of quality control materials (QCMs)*

[27] ISO/IEC Guide 98-3, *Uncertainty of measurement — Part 3: Guide to the expression of uncertainty in measurement (GUM:1995)*

[28] ISO/IEC Guide 98-4, *Uncertainty of measurement — Part 4: Role of measurement uncertainty in conformity assessment*

[29] IEC Guide 115, *Application of uncertainty of measurement to conformity assessment activities in the electrotechnical sector*

[30] *Joint BIPM, OIML, ILAC and ISO declaration on metrological traceability,* 2011 [2]

2) http://www.bipm.org/utils/common/pdf/BIPM-OIML-ILAC-ISO_joint_declaration_2011.pdf

erence materials—Good practice in using reference materials

[25] **JIS Q 0035** 標準物質—認証のための一般的及び統計的な原則

　　　注記　対応国際規格：**ISO Guide 35**, Reference materials—Guidance for characterization and assessment of homogeneity and stability

[26] **ISO Guide 80**, Guidance for the in-house preparation of quality control materials (QCMs)

[27] **ISO/IEC Guide 98-3**, Uncertainty of measurement—Part 3: Guide to the expression of uncertainty in measurement (GUM:1995)

[28] **ISO/IEC Guide 98-4**, Uncertainty of measurement—Part 4: Role of measurement uncertainty in conformity assessment

[29] **IEC Guide 115**, Application of uncertainty of measurement to conformity assessment activities in the electrotechnical sector

[30] Joint BIPM, OIML, ILAC and ISO declaration on metrological traceability, 2011 [2]

---

2)　http://www.bipm.org/utils/common/pdf/BIPM-OIML-ILAC-ISO_joint_declaration_2011.pdf

200 ISO/IEC 17025

[31] International Laboratory Accreditation Cooperation (ILAC) [3]

[32] *International vocabulary of terms in legal metrology (VIML)*, OIML V1:2013

[33] JCGM 106:2012, *Evaluation of measurement data — The role of measurement uncertainty in conformity assessment*

[34] *The Selection and Use of Reference Materials,* EEE/RM/062rev3, Eurachem [4]

[35] *SI Brochure: The International System of Units (SI)*, BIPM [5]

---

[3] http://ilac.org/

[4] https://www.eurachem.org/images/stories/Guides/pdf/EEE-RM-062rev3.pdf

[5] http://www.bipm.org/en/publications/si-brochure/

[31] International Laboratory Accreditation Cooperation (ILAC) [3]

[32] International vocabulary of terms in legal metrology (VIML), **OIML V1**:2013

[33] **JCGM 106**:2012, Evaluation of measurement data—The role of measurement uncertainty in conformity assessment

[34] The Selection and Use of Reference Materials, EEE/RM/062rev3, Eurachem [4]

[35] SI Brochure: The International System of Units (SI), BIPM [5]

---

3) http://ilac.org/

4) https://www.eurachem.org/images/stories/Guides/pdf/EEE-RM-062rev3.pdf

5) http://www.bipm.org/en/publications/si-brochure/

対訳 ISO/IEC 17025:2017（JIS Q 17025:2018）
試験所及び校正機関の能力に関する一般要求事項 ［ポケット版］
定価：本体 6,800 円（税別）

2018 年 10 月 19 日	第 1 版第 1 刷発行
2019 年 3 月 25 日	第 2 刷発行

編　　者　一般財団法人　日本規格協会

発 行 者　揖斐　敏夫

発 行 所　一般財団法人　日本規格協会

　　　　　〒 108-0073　東京都港区三田 3 丁目 13-12 三田 MT ビル
　　　　　http://www.jsa.or.jp/
　　　　　振替　00160-2-195146

印 刷 所　株式会社ディグ

© Japanese Standards Association, et al., 2018　　　Printed in Japan
ISBN978-4-542-40278-2

● 当会発行図書，海外規格のお求めは，下記をご利用ください．
　販売サービスチーム：(03)4231-8550
　書店販売：(03)4231-8553　注文 FAX：(03)4231-8665
　JSA Webdesk：https://webdesk.jsa.or.jp/